The Teacher's Guide to Scratch – Beginner

The Teacher's Guide to Scratch – Beginner is a practical guide for educators preparing beginners-level coding lessons and assignments in their K–12 classrooms. The world's largest and most active visual programming platform, Scratch helps today's schools answer the growing call to realize important learning outcomes using coding and computer science. This book illustrates the benefits and fundamental building blocks of Scratch coding, details effective pedagogical strategies and learner collaborations, and offers actionable, accessible troubleshooting tips. Geared toward the fledgling user, these four unique coding projects will provide the technical training that teachers need to feel comfortable and confident in their skills and to help instil the same feeling of accomplishment in their students. Clear goals, a comprehensive glossary, and other features ensure the project's enduring relevance as a reference work for computer science education in grade school. Thanks to Scratch's cost-effective open-source license, suitability for blended and project-based learning, notable lack of privacy or security risks, and consistency in format even amid software and interface updates, this will be an enduring practitioner manual and professional development resource for years to come.

Kai Hutchence is CEO and Founder of Massive Corporation Game Studios as well as its subdivision, Massive Learning, which focuses on educational products and services. Through his established coding support partnerships with elementary, middle and high schools, post-secondary institutions, and provincial and national organizations, Kai has taught over 20,000 students to code and over 2,000 educators to code and teach coding.

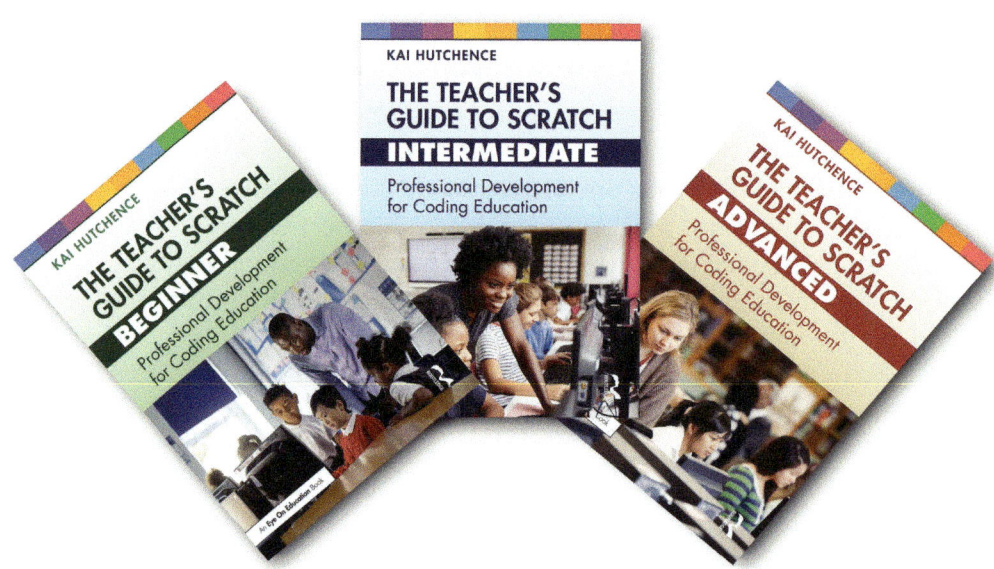

The Teacher's Guide to Scratch – Beginner

Professional Development for Coding Education

Kai Hutchence

Routledge
Taylor & Francis Group

NEW YORK AND LONDON

Designed cover image: © iStock

First published 2024
by Routledge
605 Third Avenue, New York, NY 10158

and by Routledge
4 Park Square, Milton Park, Abingdon, Oxon, OX14 4RN

Routledge is an imprint of the Taylor & Francis Group, an informa business

ISBN: 978-1-032-49908-6 (hbk)
ISBN: 978-1-032-44817-6 (pbk)
ISBN: 978-1-003-39901-8 (ebk)

DOI: 10.4324/9781003399018

Typeset in Palatino
by Apex CoVantage, LLC

Additional copyediting and developmental editing provided by Gui de Souza Rocha.

Scratch is a project of the Scratch Foundation, in collaboration with the Lifelong Kindergarten Group at the MIT Media Lab. It is available for free at https://scratch.mit.edu.

The Scratch name, Scratch logo, Scratch Day logo, Scratch Cat, and Gobo are Trademarks owned by the Scratch Team and are used for identification and do not constitute or imply ownership or endorsement by the Scratch Foundation or Lifelong Kindergarten Group at the MIT Media Lab.

With thanks to Gui de Souza Rocha and Terry Hoganson for their assistance, encouragement, editing, testing, and feedback.

Contents

Meet the Author

Born and raised in Regina, Saskatchewan, Kai Hutchence was fortunate to be the son of a math professor (and later research scientist) with an interest in computer science. With his father's help, he taught himself BASIC coding and, as a teenager, HTML, Visual Basic, and other languages. Living in a province with, at that time, almost no tech companies, Kai explored careers in politics, non-profits, and restaurateuring before moving to Ontario to work in game development.

In Ontario, he co-founded a game studio that went on to garner over four million downloads of its mobile apps. Living amidst a thriving game development community, he made connections with major players in the industry, helped aspiring creators publish their own games, and saw a community transform into an industry. After almost a decade in Ontario, he moved back to Saskatchewan to take those lessons and try to build up the industry in his home province, a lifelong goal.

In Saskatchewan, Kai launched SaskGameDev as an organization to build, nurture, and support game development in the province. Additionally, he launched his own game development company, Massive Corporation Game Studios. To support the long-term growth of game development in the province, he took up a consultancy role to help spread coding education through schools and non-profits.

While dedicating a portion of his time to teaching, Kai has helped establish coding support partnerships with elementary, middle and high schools, post-secondary institutions, and provincial and national organizations. He has helped develop coding and AI instructional material for nationwide use. Through his multiple partnerships and public profile, he has taught workshops, given lectures, provided industry mentorship and vouching, and facilitated internships with post-secondary institutes across Canada. Following up this success and newfound passion for coding education, Kai launched Massive Learning, a subdivision of Massive Corporation, to focus on educational products and services.

He currently lives in Regina, Saskatchewan. He continues to make games through Massive Corporation and provide educational services though Massive Learning. He enjoys cooking, gardening, and hiking, when he's not teaching, writing, or developing.

Foreword
by Amon Millner

I have a healthy appreciation for those who take on ambitious projects. Writing a thorough guide about an evolving programming environment, geared toward teachers who manage an ever-changing education landscape, certainly falls into that category. The result – driven, I'm sure, by the type of "hard fun" that learning visionary Seymour Papert encouraged – is a well-written resource that will serve many types of educators, creators, and learners.

As a member of the core team that designed and delivered the first version of Scratch, I have made it a mission to create resources for those same groups. The journey that led me to being a part of the Scratch story, which I'll share here in part, underscores why I appreciate the voice that author Kai Hutchence brings to this book and series.

Before joining the Scratch team and Mitchel Resnick's Lifelong Kindergarten Research Group (LLK) at the MIT Media Lab, I spent time mentoring at after-school learning centres called Computer Clubhouses. (Now they're called "Clubhouses", after dropping "Computer" in 2015.) The Clubhouse Network offers after-school safe spaces where imagination meets technology in areas where they are needed the most. These spaces adhere to a learning model built around four central themes: learning by designing, following your interests, building a community, and fostering respect. By the time I met Mitchel in 2002 at a Clubhouse event for teens, I had mentored at a handful of Clubhouses, from the neighbourhood in which I grew up to others that were close to the universities I attended. Mitchel and I spoke at length about how we would attempted to share our passion for programming in many Clubhouses. We lamented that the existing tools for coding were not compelling enough to gain traction in environments where young people could have fun across domains (putting posters of digital art on the walls, making original soundtracks, etc.). Many of the inaugural team members knew what made the Clubhouses special and so shaped Scratch into a tool that would thrive in them.

Scratch's online project-sharing feature was designed to be like a virtual wall that all creators in the community could post to. That way, people of different backgrounds and interests could appreciate projects ranging from animated cartoons to interactive simulations. People who identified as artists or game designers would sometimes post how-to projects to encourage others to create in those domains. Communities would grow as Scratchers shared their expertise across groups. Some Scratchers enjoyed building bridges and making connections between communities.

With all that in mind, I see in Kai Hutchence those same characteristics that some of the most prolific Scratchers embody: he is a community builder who brings people together through ambitious projects. Kai credits his experiences in disparate sectors for giving him the insights and perspectives that allowed him to write this book and series in the way he did. He has been a web/app/game developer who also taught across every tier of formal education and professional development, from primary to higher education, and has experience in entertainment industries.

Readers will find influences from each of those experiences in this book as Kai links foundational educational theorists and theory, from Dewey to constructivism, to better understand the perils and possibilities in the technology-saturated educational system of today and tomorrow. He shows deference to teachers, knowing that they are the experts of their own circumstances, and honours their expertise in his writing. The book's structure will support those accustomed to tinkering with topics as well as those who prefer planning methodically. Some of the cornerstones of Scratch are affording space to try things out, allowing for reflection, and encouraging iteration. Hutchence eloquently writes in a way that centres these same cornerstones throughout.

Educators who use this book to build up their Scratch expertise are also likely to develop keen abilities to nurture their students' Scratch skills. The diverse journeys made possible by Kai's guidance will undoubtedly span a range of topics and curricula, from computing and mathematics to social studies and art. Along these journeys, teachers, especially those with no coding experience, will encounter situations where Scratch – its tools and/or ecosystem – may seem perplexing. Hutchence deftly anticipates and plans for a variety of instances in which confusion could potentially stifle readers. He does an excellent job of explaining possible paths forward for readers who follow the exercises he proposes or when Scratch responds to programs in ways they did not expect.

Of the many connections made in this book between Clubhouses and Scratch, it warms me to see Kai Hutchence highlight how coding is more than a path to a career or professional pursuits. He makes a compelling case for creative expression and lifelong learning as critical outcomes of coding instruction. Through this ambitious undertaking, Kai has shown that he is a lifelong learner capable of inspiring others to be the same. I am confident that the type of thinking that this thorough and thoughtful guide cultivates will continue to serve teachers amidst the ongoing, and simultaneous, advancement of computing tools like Scratch and evolution of the education system at large.

Amon Millner is Associate Professor of Computing and Innovation and Director of the Extending Access to STEM Empowerment (EASE) Lab at the Olin College of Engineering, USA. A co-inventor of the Scratch programming language and former Chairman of the Board of the Clubhouse Network, Dr Millner later served as Founding Advisor of Unruly-Studios and co-invented Unruly Splats, a cross-curricular STEM learning tool that combines coding games for kids with active play. As Curriculum Director, Dr Millner teamed with Mighty Picnic and PBS Kids to launch LYLA IN THE LOOP, an animated series showcasing a family that leverages creative problem-solving and computational thinking skills to address a range of everyday problems together.

Reader's Key

Sidebars

Style Legend

To help keep the different concepts involved clear for readers, we have adopted the following text stylings to denote particular things relating to Scratch projects:

Style – meaning
Object – a sprite or the stage
●Code Category – one of the colour-coded categories of code blocks
[Code Block] or **(Code Block)** or **<Code Block>** – any one of the many code components in Scratch (the brackets help convey the shape of the code block)
"Variable Name" – a variable added to the project
//Script Name – a connected sequence of code blocks (a stack or script) in a project object

1

Why This Series?

I wrote this series to try to help address the tremendous changes happening in education as it attempts to incorporate coding into the standard curriculum. It is important that we recognize these challenges, and opportunities, and look at the broad picture before we get caught up in the fine details of an applied skill. While every era within memory has had educators struggle to adapt to new technology and changes in society, the rate and scale of change we face today is unparalleled. It is natural that educators feel overwhelmed or lost in the turmoil. The hope is that this series can offer not only theory and explanation of why these changes are necessary but also practical steps to tackle the challenge. My objective is to give some view to the future, the goals, the purpose that we are reaching for as a society, but also the grounding in applied skills that you will use to build a practice around that can help you transform your classroom into a training ground for world changers.

We may feel tremendous pressure from change, from coding to artificial intelligence, genetic sequencing and editing, climate change, digitization, overpopulation, robotics, renewable energy, resource shortages, democratization, and nanotechnology – the scale and power of change is staggering on all fronts – but we may also become some of the most justifiably proud teachers in history, leading a generation that will see some of the most radical achievements of our species: the elimination of diseases, the eradication of poverty, the elimination of toil, even extra-planetary habitation. The challenges are big, but the opportunities are vast.

DOI: 10.4324/9781003399018-1

The Digital Revolution

Tech is unavoidable these days. Over the past ten years, virtually every school has converted to digital systems: classroom management systems, learning management systems, gamification, digital labs, virtual meetings, cloud software suites, plagiarism checkers, e-learning courses, virtual workshops, computer carts, and student log-ins for school networks. Tech has completely infiltrated the modern classroom; it is silly to think that the skills to build, manage, or edit all these tools would not also need to become a part of the curriculum. We are seeing, slowly, every jurisdiction roll out new coding requirements in their curriculum. As hard as it might seem to adapt, I think this transition is at least a generation late. We have known computers were the future since the space race. We have been negligent to have not taken the steps earlier to make this process easier. But we cannot change the past, only move forward.

A lot of John Dewey's thoughts on education are still surprisingly relevant a hundred years after he wrote them. We are still facing an uneasy crossing-over from traditional to progressive learning. We have seen multiple schools of progressive learning begin to coalesce, and looking at various jurisdictions, we can see different approaches being adapted as standard. Inquiry-based learning, project-based learning, game-based learning, personalized learning, differentiated learning, flipped classrooms, kinaesthetic learning, direct instruction – it is a maze of models and buzzwords. In truth, all styles have value, with their own strengths and weaknesses. If there were a single correct model, we would all be using it by now. We cannot assume there will be a single winning way to transform education or incorporate coding into the classroom. It will need to be done in a way that suits your jurisdiction's dictates, your school's community, and your classroom's needs. It is precisely this diversity of needs and styles that makes it so hard to adapt. Teachers cannot simply copy and paste a solution; it does have to be adapted, and that does mean time and effort, not exactly things we have a lot of to spare.

To do this, teachers need to understand both the tools and theory of the subject and have an understanding of the pedagogy around it. It is a lot to understand; it is all foreign and new. It is too much to be a short or simple process. It is also far too much to cover in one book. Step 1 is to understand the subject and the tools to explore it. That is the goal of this book: provide a clear, deliberate plan to master the tools so that you have the applied knowledge to deal with the subject. Then hopefully, if this is successful, I hope to follow up with guidance on the larger picture of pedagogy and learning theory for the bigger how and why we need coding education and practical advice on how to pull it off.

The Problems

Change is never easy. There are significant challenges to overcome, some from the nature of the change, some from the nature of the system, and plenty just from the situation we find ourselves in. The reason a book like this is necessary is the lack of training for teachers to understand coding. Our education faculties should have adapted at any point in the last 60 years to begin this process but largely have not. Teacher training should have been working us toward a greater understanding, slowly but surely, for generations. Then this would not be so foreign. We would not lack experts within the education system if teachers had been exposed to this sooner and under less-intensive demands. It could have been a matter of exploration and inquiry instead of a matter of compliance and demand.

We have failed to provide preparation, staged learning, and incorporation around coding. Now teachers across the world are being faced with an immediate need for adoption. Unfortunately, teaching continues to be a demanding, life-consuming job. If they could not get trained before going into the classroom, when, pray tell, are they supposed to manage to learn and adapt to this new field of knowledge? There is no time off for it scheduled; I have not heard of many paid sabbaticals to train in it and become comfortable with it, or to make plans for their classroom to adopt it. I have not heard of anything else being taken off teachers' plates to accommodate for the new demand. Seemingly, most jurisdictions and school boards, like economists, have not learned to subtract. What time and support is being offered to teachers to make this change? This change is far greater than anything else we have thrown at teachers.

When we look past the aforementioned two systemic/governmental issues, there is still plenty of hurdles. While coding is being taken up and explored by the education community, it is still quite new. There are resources, but they are still limited. There are lots of singular projects available to see, study, or run, but there are few comprehensive guides and supports. Teachers need learning pathways through this difficult new field but are largely at the mercy of fate, simply searching for one shot and hoping they will be appropriate and complementary, often with little understanding of what is needed or possible.

Thankfully, the coding educational tool Scratch is a leading method for tackling coding, and it does have a lot of resources available, and some community built around it. So one can find lots of projects to work with. But even with this great tool, there is little in the way of more comprehensive supports. This book is largely about that, not just teaching projects, but teaching underlying theories and professional practices through projects. Without understanding the theories and principles underlying code, we are really just falling back to rote practice and repetition, having our students do but not truly knowing or understanding.

One of the reasons it has been so difficult to incorporate coding into education is that so few people have dealt with both spheres of knowledge: professional experience as both educators and as tech professionals, especially programmers. This means very few people can act as savants and guides to translate between these two fields. Education tends to be a very extroverted pursuit; programming a very introverted pursuit. The training to be a professional in either field tends to be fairly significant. There is also very little overlap in the skill sets, theories, and references. So this is not likely to resolve easily either. Without tech people that know and engage in educational practice, and without education people that know and engage in tech (at both the creation and maintenance level), we will have few luminaries to guide the processes of this revolution.

The Good

We can be honest about the challenges and the problems but shouldn't let that stop us or trap us in negative thinking. There are challenges to overcome for sure, but there is a lot of good that this work will achieve. The goal of coding integration in classrooms, and of tech skill development in schools, is one that will have tremendous positive impacts. So how do we acknowledge the challenge, act on our realities, and maintain a growth mindset?

The fact is that we need to revolutionize education because there are revolutionary new capabilities at our fingertips. The world has changed, and we have to react to that change. The amazing technology that has transformed our world has also provided us tools for that transformation. We have already adopted to all kinds of classroom tech (EdTech); coding is simply the most fundamental and the biggest step we have to take. Thankfully, tools exist to make this transition happen, communities exist to support the change, and individuals and organizations are out there fighting for the change and to support it. You might feel like you are alone in this, but you are not; you just might not have found or recognized your allies yet.

This book series is focused on one particular tool and its associated communities. Scratch, or more specifically, Scratch 3, is an amazing tool that can help us achieve this revolution, at least for a good chunk of basic education. Available free online at http://scratch.mit.edu as either an online tool or as an offline downloaded program, it can allow teachers and students to learn a wonderful array of coding knowledge while exploring a wide range of

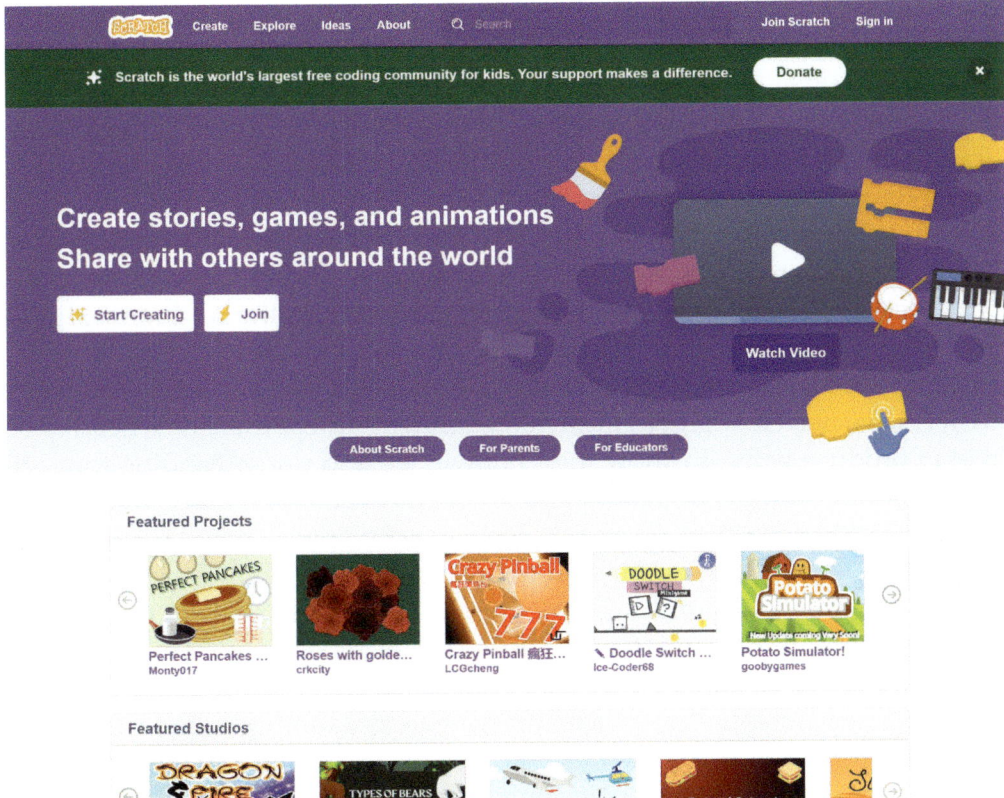

Figure 1.1 The http://scratch.mit.edu home page.

creative possibilities. I think Scratch is the best tool to support this change in education, built from the ground up to provide accessible, interesting, easy, but deep coding educational experiences.

In teaching coding, we will be providing our students with a tremendous benefit. Students need to be exposed to coding and understand it at least at a basic level for the world of tomorrow. While few people would question coding as a skill today, we always need to remember that we are trying to prepare students for the world where they will be adults, citizens, and professionals in ten years or more. We have to be forward-looking, and coding is a must for that future. It is already a dominant influence on our lives, and it is only getting stronger. Code is the technological tool this generation needs to understand to engage, control, build, or invent the world of tomorrow.

Coding is more than a career, though. Through teaching coding, we teach our students new ways of thinking. Whether they pursue it as a calling, enjoy it as a hobby, or just have it as part of their skill set, coding creates lifelong learners. Code is always changing and is so deep in capability that it can never be fully known. To be a coder is to be a learner. There is always more to know and try; learning to learn is fundamental to coding, and perhaps it is the best subject to instil that drive to learn and to be effective learners.

Coding is also very importantly a creative practice. Coders create things. Even if we did not need to teach code, the world that technology has created demands a change in teaching. Answers are at our fingertips. The vast sea of knowledge humanity has accumulated is available and searchable like never before, and our teaching practice must change as a result. We cannot simply check if students can get the answer; simple recitation of knowledge is easily replicated. While we still want informed and knowledgeable students, and they should learn rather than always be dependent on easily searchable answers, we need to adapt to this reality. Students, more than ever, need to be able to produce work, to show their knowledge and understanding through application. Coding is a great venue for them to show off their knowledge, interests, and capabilities. We can go beyond answers and get to applications. We know we need to move beyond traditional testing methods and teaching to them. Coding is an amazing way to provide not just skills of its own but also a medium for showing their work and comprehension of any subject.

We also need to step back and be realistic about what we are doing. The capability and potential are a vast, inspiring, and daunting vista to behold. It is easy for people to get discouraged or feel lost with such a challenge, so it is important we remember our own context. No one person trained the world's greatest writers or dancers. They started by learning their ABCs or how to walk. The next person taught them words or left and right. It is all iterative. It is a long process. We do not need to feel like we have to master the whole depth and breadth of the subject. We do not have to feel like we personally can take our students anywhere. Your classroom is more a guidepost they can aim for and be aimed at the previous guidepost from, not a comprehensive street map of the whole world of possibilities. You will offer your students some answers and opportunities and then point them onward to others. This revolution will take time; the need is great, but it is not just on you, not just on this year's classes. It is a staged, staggered, spread-out adoption. You can pace yourself, and you should.

Teachers have always had a tremendous opportunity to shape society and the future. With coding, we are living up to the practice of thousands of years, preparing our students for the world they will live in. We may be treading new ground, but we are also doing exactly the same thing every teacher ever has. We are going to challenge ourselves and grow personally and professionally by taking on the challenge of learning coding. In doing that, we will provide our students the best opportunities we can for their future.

What This Book Series Is

This series is your all-in-one professional development training course for coding in Scratch. My approach to this comes from having been a part of three worlds that I think combine to provide me with some unique insight to help with this revolution in education. I have professionally worked in tech as a web, app, and game developer, which give me a range of deep technical knowledge of coding and other tech skills. I have worked in education as a coding instructor for years in primary, secondary, and post-secondary education. I have taught coding and game development to thousands of students and educators of all ages and have worked across the nation, seeing all kinds of classrooms, teachers, curriculum, and standards. I have also been an entertainer, creating both tabletop and video games to entertain the masses.

Because of this combination of skills, I think I have a fairly unique perspective to tackle the challenge of helping teachers incorporate coding into their professional toolkit. I try to combine the technical skills with the awareness of teaching at the practical level while also making it all fun and appealing. My hope is to give you the technical skills to be a star, the teaching insights to give you a leg up on handling the challenge, and fun examples that will engage you and your students and help them be inspired to create.

This first book in the series is written for beginners. We are starting from nothing and teaching everything you will need to know to get started. We will introduce all the components of Scratch, where to find them, and how to use them. We provide a series of projects that can take someone from no experience with coding through to producing their own simple projects and, importantly, inspire them to create. We have a follow-up book for intermediate skills, and another for advanced, so you will be able to use these books to grow your skills from nothing right up to the limits of Scratch's capabilities. Let us take a look at the different sections of this first book and what they will cover.

The Why of Coding Education

The first three chapters of the book ("Why This Series?" "Why Coding?" and "Why Scratch?") are about why we need coding education and why Scratch is the tool I recommend. In these three chapters, we will explore the concepts surrounding coding education, look at the challenges and opportunities, and examine how Scratch in particular is a great solution to address these issues.

Figure 1.2 *The Teacher's Guide to Scratch* book series.

Scratch Basics

This large chapter is a whirlwind tour of Scratch. It takes you through every aspect of the system so you can understand its components, as well as the structural whole of how Scratch works. This chapter will familiarize you with all the different tools Scratch provides, how to navigate around it, and how to achieve all the basic tasks with it.

Beginner Projects

This book has four projects in total to introduce Scratch and develop your digital skills with it. These chapters provide you with a simple project suitable for early to mid-grades (approximately grades 1 to 4) that will provide you with a fun project while introducing some of the basic tools and concepts. At the end of the four beginner projects, we have a check-in chapter, where you can review what you have learned through the beginner projects. We also offer some suggestions for more beginner-level projects to explore on your own if you are not ready to advance to our next book on intermediate-level Scratch just yet.

Follow-Up

In this chapter, you'll be provided new challenges to improve the projects presented in this book, refining them, adding features (initially cut for simplicity and brevity). There are numerous recommendations for expansions and revisions to all five projects in the book that you can challenge yourself

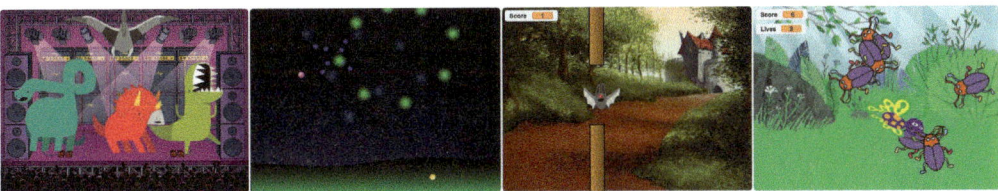

Figure 1.3 The four beginner projects covered in this book.

or your students with. We will cover even more ideas for refinements and expansions in our intermediate and advanced Scratch book that can prompt a return to this simpler projects with the knowledge gained through those other projects.

Troubleshooting

This chapter provides you some key advice on how to handle the problems that will show up while teaching coding with Scratch. You will find some insight on how to handle bugs in the classroom, as well as exact ideas and approaches to take with the most common issues I have seen in teaching thousands of kids and adults with Scratch for over six years.

Glossary

Lastly, you will find a reference for any particular terms used in the book that might need some clarification.

2

Why Coding?

For most educators, there probably isn't much option on this. Coding is being added as a curriculum requirement for schools in most jurisdictions. But we don't need to have an attitude of resignation about it. It is a good thing, and hopefully through this book you'll come to see not only the benefit to our students for learning coding but also how it can be a fun and engaging tool you can enjoy using in your classroom paired with other subjects.

There are seven areas that I see as being key reasons we should embrace coding in education. Some of these are more speculative, looking to the future; others more fundamental leaning on proven value. Some are specific to coding and technology, while some have benefits to everyday life and complement other subjects and skills.

Digital Careers

The most obvious benefit is that kids that know how to code can code. Programming is an in-demand career, and often high-paying. By teaching more kids to code, we can provide more kids the career path into a programming career. However, the benefits go far beyond programming. We say *digital careers* and not *tech careers* because of how widespread the need is and how widespread the skills are. By engaging our students in coding, particularly with Scratch, we aren't just setting them up to be programmers alone but to understand and engage with technology and digital skills broadly. They

DOI: 10.4324/9781003399018-2

might become programmers, but our goal isn't to just produce programmers, because no society can function with everyone going into one career. Likewise, we don't teach health so everyone will become a doctor. But by understanding and engaging with digital tools, they will more easily find their way into all kinds of digital careers, from programmers to web developers, to graphic designers, to cybersecurity specialists or network technicians. There are technical, creative, management, sales, social, and even educational-related jobs in the digital space; we open the doors wide to these new careers through coding.

Automation

Technology is rapidly changing the world, in many ways for the good, in some ways for the worse, and in a lot of cases, we don't know how yet. Automation is going to be the biggest story in labour and economics in the 21st century. Artificial intelligence (AI) and robotics are advancing to a degree where large-scale job replacement is an undeniable reality we'll be facing in rapid order. We don't understand what this is going to do to society, what good and what bad it will cause. It can, and should, mean a higher quality of life for people and more leisure and quality time as we put menial tasks behind us. But it could mean mass unemployment and social unrest, like the Industrial Revolution did. About our only strategy for now is ensuring kids have future-ready skill sets. If automation is the next revolution, then understanding the tools of that revolution will hopefully give them better options to handle that change. Of course, we also want to make sure they have the grounding in social sciences to understand what (political) policies we need to ensure positive outcomes of these changes too.

Sociological

Why do we teach small engine repair in high school? How many people does lawnmower repair employ? We teach small engine repair not because we want a society that can repair their own small engines but because it explains a fundamental technology that our society is built on. There's a good chance you'll never need to repair an engine yourself, but getting hands-on with them and understanding how they work prepare you in some way for being a citizen in our engine-driven world. You will encounter this technology. You will own this technology. You should have some grasp of how it works. As we move to a digital society, it's important that an informed citizen has a basic grasp of how this fundamental technology works. Even if they don't become professionals dealing directly with it, the widespread knowledge empowers our society. We create a literacy of the components, methods, and capabilities so that our students will be informed citizens. As digital policy becomes more and more critical, we need the population (and lawmakers) to understand the methods, opportunities, and dangers of this technology. By laying a groundwork through this basic engagement and education, we can be best prepared for the societal needs surrounding this critical technology and related infrastructure.

Opportunities

Have you ever thought about being an inventor? Can you imagine what it would be like to be the person that invented the lightbulb? The electric oven? The steam train? We think back on previous technological revolutions with wonder, awe, and often, a touch of jealousy. The answers seem so obvious now – couldn't you have made that leap if you had been in the right place at the right time to seize the day at the dawn of a new technology? Digital technology, especially AI and robotics, is like that, right now. One of our students could be the next great inventor. This is the right time and place to do amazing things. Teaching them the fundamentals to understand this new era of opportunity will help them bring their ideas to life and possibly change the world. By making coding available to all students, we can hope to overcome biases in these amazing opportunities to create a more equitable world.

Creativity

Coding often has a perception as a very math- and logic-focused activity, but I think the reality is quite different. While our strong math students tend to easily start working in coding, it is a far more diverse subject and opportunity, and we should embrace its myriad uses and interests. Coding isn't about the code. We don't code to admire our code. We code to build things. Coding is a tool that allows us to create.

We can create just about anything with code and certainly involve any topic or subject. Music videos, interactive novels, video games, simulations, poetry and visual art experiences, slide shows, interactive recipes, and more – there is so much we can do through coding. We can explore whatever interests we have and bring them to life. We can express ourselves and share our ideas and creations with the world. Whatever ideas or interests we have, they can be explored through coding. Through that exploration, and through those projects, we get challenged to learn and think to successfully build things. Our own interests can drive us to engage and succeed, and we learn not just coding but processes of design, work, evaluation, revision, and research. Coding is the medium, but through our creative pursuits, we can use it to learn and master any number of skills and fields of knowledge.

Coding allows us to explore our creativity in new ways, opening new possibilities and dimensions for our ingenuity. Coding tools may give us access to things we didn't have access to before, such as digital music composition, digital art, or animation systems. They may allow us to explore things in new ways, such as simulation, AI-controlled agents, randomization, procedural generation, data visualization, and large-scale data analysis. Coding can reveal emergent systems that show us things we didn't perceive or appreciate until we could see them in explorable, pausable, rewindable, recreatable digital simulations. Through digital systems and coding, our creativity expands to include not just new means, new motivations, and new types of creation but new ways to explore, understand, analyze, appreciate, and share them all.

Thinking

Ideally, we don't just inform our students; instead, we teach them to think and learn for themselves. Coding is an amazing tool for this. Critical thinking is built on logic, just like coding. When students engage with coding, they inescapably engage with logic. But because coding comes to life and a computer interprets their code, they get instant autonomous feedback on their logic. Coding acts as an active and engaging practice of logic. The coder writes their instructions, the computer reacts to them, and then they must interpret how the computer reacted to them to understand if their code was right, or where and how it went wrong. Coding can be play, and it can be used for entertainment, but it is also deeply analytical in nature.

Beyond critical thinking, coding helps us explore the concept of "design thinking". Coding is about building things, and that process can be encapsulated by design thinking's five-step process: empathize, define, ideate, prototype, and test. Whether we recognize this process or not, as coders we go through it inherently, but we can also formalize it and elucidate this thinking and process. It can help highlight the social nature of a project, ensuring we think from the user's perspective, not only logical thinking, but empathetic thinking as well.

Most commonly, coding is cited as the way to introduce "computational thinking" to students. Computational thinking isn't about thinking like a computer but rather about problem-solving, which is what we try to use when computing and coding. Computational thinking is built around the concepts of decomposition, abstraction, pattern recognition, algorithms, modelling and simulation, and evaluation. Through these six concepts, we can break down problems and create and implement solutions. Decomposition breaks problems down into simpler components. Abstraction turns specifics into generalized concepts and processes. Pattern recognition allows us to reuse and simplify concepts and processes. Algorithms build repeatable processes to address issues. Modelling and simulation allow us to build representations of problems and solutions and see them in action and interaction. Evaluation allows us to access our processes and solutions to see if they are adequate or if additional actions are needed. These concepts don't just apply to coding; they are easily explored and reinforced by the practice of coding. Through coding, our students can learn new ways of thinking, with projects that give direct and independent feedback for them to correct and refine in response to. With coding, a computer turns into a logic tutor. By exercising these skills and working with logic, we can expand their critical thinking skills and give them a hobby or career that will keep honing those critical thinking skills.

Engaging

I don't want to shock anyone, but kids, it turns out, like video games. There are plenty of reasons teachers might not like them as much, but coding is a great way that we can engage students and, through it, create healthier relationships with digital entertainment. I, too, worry about excessive screen time, especially for the very young. I, too, worry about bad habits, and there are definitely digital products I would not use or recommend children use. But we do need to have a non-hysterical response to whatever problems there might be with digital entertainment. Moral panic is not an appropriate, critical thinking–backed response. Video games are the dominant form of entertainment for the younger generation. For the most part, it's replacing a lot of TV watching time. With video games, we at least get some amount of thinking, strategizing, socializing, and reacting. A precious few even include physical activity. Video games are not a pure bad or pure good influence. With thoughtful engagement, we can try to minimize and avoid issues while still utilizing their positive benefits of increased strategic thinking, analytical thinking, social bonding, creativity, and opportunity to learn.

Students are very engaged with video games, and that is an easy win for getting them interested in coding. It's pretty rare to see a student who doesn't have some visible excitement when they're told that they'll get to make their own games. The key thing to remember is that we are changing the relationship they have with video games. By engaging them with coding, we turn them from consumers into creators. This doesn't just mean they can spend time creating games; it also changes how they play them. They'll be more inspired by them, engage in more critical thinking about them, analyze them to understand their processes, see themselves as creators, and engage in dreaming up new areas, stories, characters, and more. It changes their relationship with the medium and, I think, greatly improves habits and understanding. Beyond the benefits of play, well documented by researchers such as Dr James Paul Gee and Dr Rachel Kowert, game design and development open players' minds to additional layers of scrutiny and reflection – thinking about the how and why of the game. For budding designers and developers, play is also review and analysis to inspire and inform their future creations. We gain all the interest and drive of engagement that video games provide in teaching coding, and all the benefits it has for learning, but we also, I think, improve a lot of the worries and possible downsides in their entertainment consumption.

Engagement with coding, of course, goes beyond just the obvious ties to video games, though. Computers are a fundamental part of every career now. Fashion uses digital design tools for creating 3D models, for planning out robotic pattern cutting, for logistics management. Athletics uses computerized training regimes, diet planning, statistical analysis, and physical technique analysis. Farming uses ever-increasing amounts of automation as we increase robotics and sensing data. Nothing escapes technology. Whatever a kid's interests are, there are ways coding is transforming that interest, whether craft, art, sports, or profession. If we understand the computer systems changing the world, we can engage our students with coding challenges that suit their interests and reveal the astounding opportunities and challenges that exist in the world. Coding is a tool they can engage with for any interest and to change the world. Coding is perhaps the most flexible subject we can teach in terms of its ability to suit the student and find a passion and drive in them to explore, create, and learn.

3

Why Scratch?

By this point we've made our arguments why we need to teach coding in schools and why educators need to know how to code. But why Scratch? Why is it the right tool for the job? There are dozens of computer languages and coding websites, so what makes Scratch our choice as the best tool for educators to work with teaching this critical new skill? Let's take a look at some of the great features of Scratch and why I think it wins as the best way to teach coding, hands down.

Accessibility and Approachability

Any tool is only useful if it can be used. Scratch has always had a major focus on accessibility and approachability. Education has had a lot of attention on accessibility lately, but *approachability* might be a newer term for some of you, as it comes from the product design world. Accessibility is all about ensuring that the design and approaches we use are suitable for our audience and allow them to use and engage fairly. Approachability is about how the audience, or user, reacts to the product. Can they understand it, do they see its value and potential, do they want to use it, or are they repelled by its complexity, abstraction, or obfuscation? When we create products or processes that are both accessible and approachable, we have a winning design to serve the widest audience best. There are a few reasons Scratch has a winning design for both accessibility and approachability.

DOI: 10.4324/9781003399018-3

Non-Commercial

A huge factor for any educator is cost. Scratch is completely free and open-source. There are no financial burdens or restrictions on use. Many other coding platforms can have a onetime or even yearly or monthly per-user cost. Scratch is the cheapest option out there. There are no contracts, no restrictions or terms on use stopping educators. There are other free tools, but as the Massachusetts Institute of Technology (MIT) developed Scratch and the MIT Creative Commons license (a legal standard for creating open, free, and shareable intellectual works), it's a pretty solid bet MIT is in this for the long haul to keep this tool free forever, and the Scratch Foundation (the non-profit that maintains it) doesn't have to worry about any shareholders pressuring them to monetize the platform.

Language Support

Scratch is amazingly available in (at the time of writing this) 74 languages! It's one of the most well-supported and accessible tools I've ever seen in tech. This doesn't just mean that classrooms all over the world can be using it; it also makes it more available for our own students who have a different mother tongue. Language in Scratch can be switched on the fly by clicking on the Settings button and then selecting Languages, or in offline Scratch, the Globe icon at the top left of the editor screen. In two seconds, you can switch to another language and all the code blocks and editor will be running in that language. It's so fast and easy you can switch back and forth to learn new words, or have ESL students quickly check to better understand something they're unfamiliar with.

Block Coding

One of the great ways to introduce coding to younger students is the concept called "block coding". A block coding system does away with the usual typed-out code, instead using premade "blocks" for users to build their code with. Each code block is a preset instruction for the computer. By connecting the code blocks, the user can create sequences of instructions for the computer, just like typed code. Different blocks connect in different ways, and some customization options, like typing in the occasional key value, allow code blocks to be almost as versatile as typed code. The important thing is that by removing typing from the process, coding becomes enormously easier to introduce in elementary/primary schools. The process of coding becomes a block-building exercise that highlights the logic structures of code and eliminates the frustration and errors of typing. This revolution in coding has opened the door for coding education like nothing else.

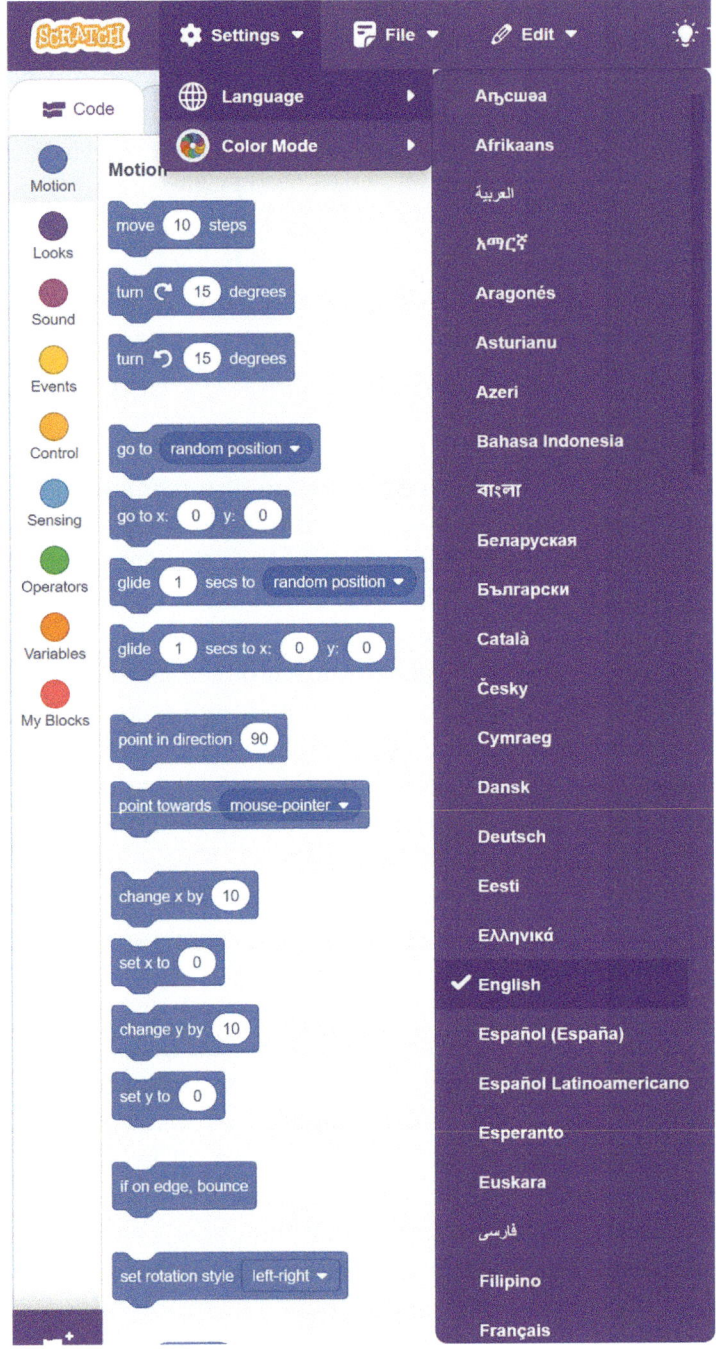

Figure 3.1 Some of the many language options for the Scratch editor available by clicking the Globe icon in the top left of the editor.

Visual

Beyond simply eliminating the hassle of typing, block coding offers a lot of other benefits. First off, its visual characteristics. It turns the labyrinthine complexity of typed code into simple blocks. Turning the confused mess of commands into singular distinct blocks, kids can see where they start and where they end, what they ask for and how they connect to other blocks. It highlights all this information, delineating use and requirements at a casual glance in a way that text code doesn't. Each code block has its own distinct piece, delineating the process into digestible steps. The relationship of code blocks by shape, colour, and position reveals key aspects of logic and interoperability. With even just a little experience, kids can start grasping computer science fundamentals simply through the recognition of shapes and patterns. The opportunity for visual learning with code blocks allows a wonderful, easy entry into coding. While highlighting fundamentals, they will grasp quicker and deeper than do learners that jump straight into text coding alone, even at higher levels.

Simple

Part of this ability to learn is that block coding also simplifies things. It isn't just that they're visual but that they are made standard; they're colour-coded, and they have expected shapes and relations. None of these things are inherently obvious in text code. But by making those things standard aspects and simplifying methods to set and consistent code blocks, it helps reinforce learning. Kids don't face the same white-page syndrome of not knowing what is possible or where to start; they have set code blocks right there for them to play/work with. They don't have to worry what all the possible options, arguments, or alternatives are.

Tactile

Lastly – and this may sound strange for anyone that hasn't used Scratch before – it's a very tactile system. That is a strange thing to say about a digital tool, but when you try it, you'll see what I mean. The way you interact with blocks, build scripts, and position sprites, it's all very visceral, so tactile it feels like you're building something with your hands. You can feel what you're doing. This level of feedback really helps reinforce actions that often in digital systems are easily missed. They help give positive feedback, build expectations, and reinforce learning through subtle but helpful clues. Shadows and outlines giving previews of actions, popping sound effects highlighting the connecting of blocks. They're all masterfully done in Scratch and help create a more complete experience for students. It makes the digital

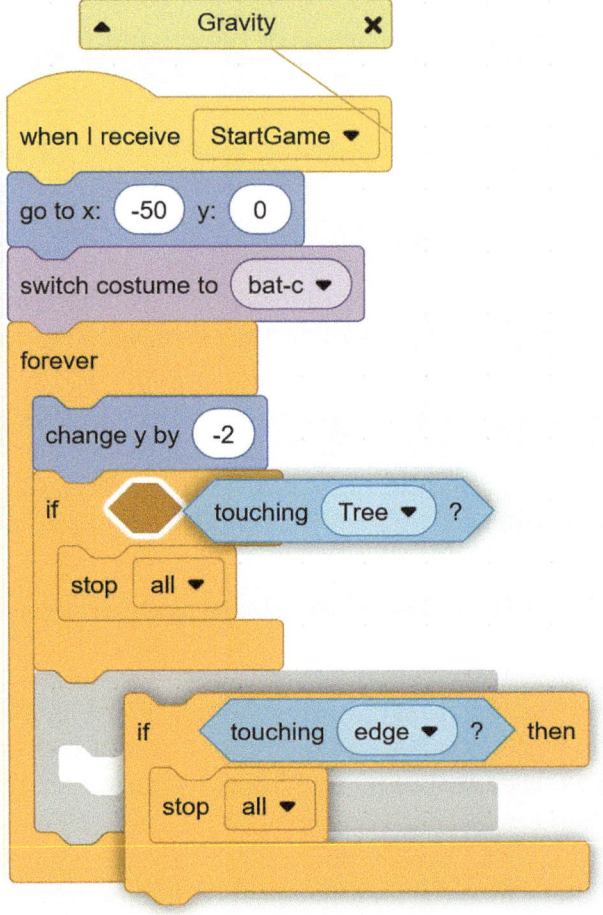

Figure 3.2 Two examples of Scratch's visual/tactile systems. Here, the white outline shows where a Touching Boolean block will insert into an If block if released. Below an If is given a shadow position at the bottom of the Forever block to indicate where it will insert into the existing code.

tangible and reinforces their influence and control of it. It helps them achieve what they want and rewards and assists their agency as they become creators of worlds.

Modes of Access

Scratch was a real saving grace for education during the pandemic. As schooling was turned upside down with learning from home, a lot of educational practice was thrown for a loop. Thankfully, Scratch was one of the real bright spots. With Scratch designed to work through any Internet browser, kids that could connect to their online classes also had the ability to connect to Scratch. But it was even more resilient than just that. Scratch also has an offline version, so users can download and install a program that they can use when they can't connect to the Internet. It can work on Windows, Chromebooks, Macs, or tablets, so the random mix of devices available to students working at home was less of a barrier than other systems. Because of the free, no-limits use, there wasn't any worry about student accounts or school licenses; any student could simply log in to any account they made in school or make new ones as they needed. While we do still face the barrier of the digital divide (students without at-home access to computers and/or Internet access), Scratch does everything in its power to be as accessible as it can be. Combined with school districts lending devices, library devices, or non-profits working to bridge the digital divide, we saw a lot of potential for access, and with planning and adaption, we can hopefully close the digital divide soon and ensure digital skills access to all students.

Audio-Visual Feedback

As I mentioned earlier, a great advantage to Scratch is how well-built the whole system is with both colourful, friendly graphics and audio-visual feedback. Scratch is (mostly) high-contrast, colourful, and colour-coded, and most of the controls give feedback. This creates a lot of feedback and engagement to the user, which not only grabs kids' attention but also clarifies information for them. While students that need to avoid stimulation might be a concern, you can try to get them used to the editor with shorter exposure times until they understand it. Once it isn't so new and confusing, the project stimulation can be adapted to the user's level, as long as educators are willing and able to adapt. One can skip using sounds (generally a good idea in school, anyways), allow for simpler graphics, lower the saturation on visuals used, tone down animations and visual changes, and such things to adapt to the level required. So we can use all the capabilities of Scratch to engage, but we can also scale things to suit the student or class.

Capability

So we know that Scratch is a great choice for logistical reasons, but what does it let us accomplish? There are a great number of options for coding education, so what makes Scratch the best choice in terms of learning outcomes, coverage of topics, or scope of projects? Again, here there's a lot of reasons Scratch is my recommendation as the best tool for coding education.

Computer Science Fundamentals

While Scratch is a simplified and approachable system, it still provides an enormous amount of the potentials of a programming language and allows exploring a large amount of fundamental computer science concepts. It does not cover everything, but it covers most of the subject that any student would likely to encounter before university.

As a tool targeting the 8–16 age range, it does a wonderful job at covering the grade-relevant concepts as we move to incorporate computer science in basic education. One of the reasons I love computer science is that it combines mathematics, logic, and information theory, some of the fundamental laws that govern the universe. When we get comfortable with these concepts, we can better understand the principles and methods that underlie all creation. We don't have to be explicit with them, but familiarizing students with them, letting them work with them and build with them, builds a tremendous foundation for higher learning. Simply by doing, students will develop critical thinking skills, because they're having to solve the problem of building what they imagined. They'll go through the process of decomposition, abstraction, pattern finding, algorithm building, modelling and simulation, and evaluation. These computational thinking practices are inescapable as students work with Scratch. Because we give them a tool for creation and self-expression, they'll be able to see the value of such a tool and then find their own inspiration to engage with it and develop their skills. We have a platform to engage students that will lead them to critical thinking and computational thinking. Through practice and discovery, they'll learn the basics of computer science even if we don't elucidate them, or even if we don't know them ourselves! You're going to learn, and you're going to like it – honest!

Digital Art

One of Scratch's big advantages over competitors is the way it incorporates art. Many coding systems either have a set library of art that users can work with, don't incorporate any art easily, or only allow uploading already-created art. Scratch not only has a set library of art and allows uploads but also goes a huge step further in having a full digital art creation system built right in. This is a monumental advantage, although it's easily overlooked by many.

The greatest advantage we can have in education is a student that wants to participate and learn. Self-motivation is probably the biggest factor of success in school, but it can be a big challenge helping students see the value in lessons and topics for them to have that self-motivation. It's great that Scratch has a library of ready-to-use art, and it's convenient that we can upload art too, but creating art is what unlocks any potential. Properly introducing students

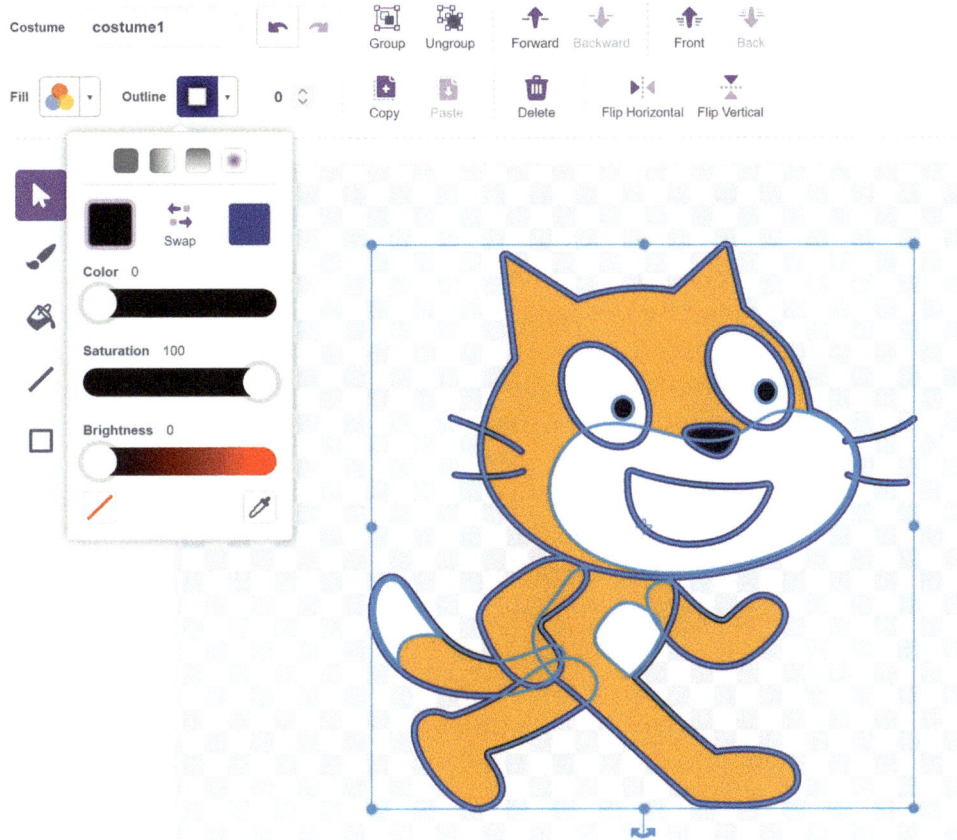

Figure 3.3 Scratch's built-in art editor.

to the art tools can unlock the full potential of Scratch. It isn't limited to just the choices provided or what they can find already made; it allows them to create anything instead. They can make art to create any kind of project, any scene, any character, in any style that they can imagine.

By having this freedom, we can support, or challenge, our students to create anything they imagine. We can tap into that yearning to express themselves, to create the world they imagine. We can tie into any interest they have and use that to show the potential and value of participating and eventually mastering the skills and knowledge involved.

Scratch's art tools support both vector and pixel art, which I'll explain later, but this comprehensive tool provides whatever we need to create what we want. They can learn not just programming but also digital art and graphic design skills that can offer fantastic career opportunities as well as self-fulfilment opportunities. Because the tools are built right into Scratch, we can seamlessly incorporate these skills and show how fundamentally tied together art and programming are and create a perfect STEAM (science, technology, engineering, art, and math) approach to learning and doing.

Extensions

Another great reason to use Scratch is its extensions. As of the Scratch 3.0 release, these optional expansions of its capabilities enable musical composition, run-time drawing, webcam video input, text-to-speech capability, text translation in project, or incorporate a number of different robotics kits. This can be great for extending the capability of Scratch to tie in with other interests of students or opportunities at schools and libraries. The Scratch team allows suggestions and feedback for new features and also runs the http://lab.scratch.mit.edu website, where they have unreleased features being tested to try out and give feedback on. Currently, there is a face-sensing AI module and an animated text module being tested that we will likely see incorporated in a Scratch 4.0 release sometime in the future. So we know that Scratch coding will allow students to work with other opportunities like robotics kits, and that the system is still continually being improved and supported.

Powerful Backing

Reliability is an important consideration when choosing to incorporate a third-party system into education. One of the reasons Scratch has caught on is the reputation of its backers. Scratch was developed at the MIT Media Lab and is still maintained and developed at MIT. Having one of the world's leading academic institutes as both the source and support for the platform is definitely reassuring. The Scratch Foundation, a non-profit founded to oversee the initiative, ensures the project has a community-minded approach and

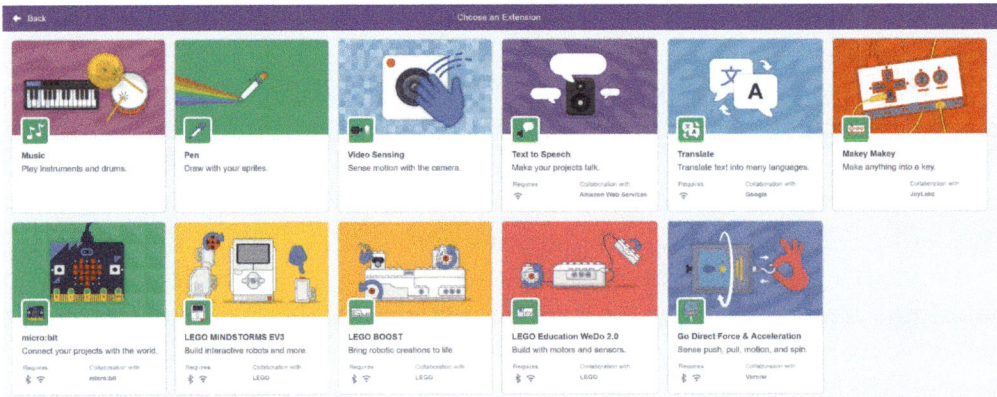

Figure 3.4 The built-in extensions for Scratch, including Music, Pen, Video Sensing, Translate, and various hardware integrations.

that it uses legal licenses that serve all of humanity. The project is maintained open-source, with its code freely accessible through GitHub (an online code repository site) for programmers that want to understand how it works or try build off it. With one of the foremost authorities on computer science helping back and develop the system, we can feel confident that it has the brains and backing to be the best tool for teaching computer science.

In addition to MIT's central role, Harvard's Graduate Studies in Education department has taken a lead role in building the education community around the tool. ScratchEd is a website and online community they built to allow educators to gather, communicate, and share best practices and lessons using Scratch. You can check out the website at: http://scratched.gse.harvard.edu. A tool that MIT and Harvard actively work to support as the means to this educational revolution is hard to ignore.

Community

I often tell students that one of my favourite things about Scratch is that it isn't just a tool for creation but also a platform for sharing. Although there's plenty to worry about with any social media, there's a lot of upsides to the way that Scratch has built the community side of the system. With the online community having been around for years now, we can get a fairly good sense of how it's worked out and can take a look at exactly what we're getting into. The Scratch team has done a good job of creating a well-behaved community that allows some level of sharing and communication but has managed to avoid the trolling, negativity, and toxicity we see in most online communities, through both passive security measures and active community policing and support.

Inspiration

We've all seen, or experienced ourselves, the creative block of staring at a blank page or screen. Sometimes we need inspiration, a hint, a clue, a spark to get us started on creating something. This is one of the most engaging sides of Scratch. With any project on Scratch, the creator can choose to share it. When shared, a project becomes public, and anyone in the world can find and try it. The Explore page of Scratch can let visitors search through Scratch projects to find games, animations, simulations, slide shows, and more of things that might interest them based on search words. In addition, Scratch highlights some of the most notable projects being created on the main page: http://scratch.mit.edu. Every time we visit, we can get a glimpse of the amazing creativity happening across the world on the platform. We can get introduced

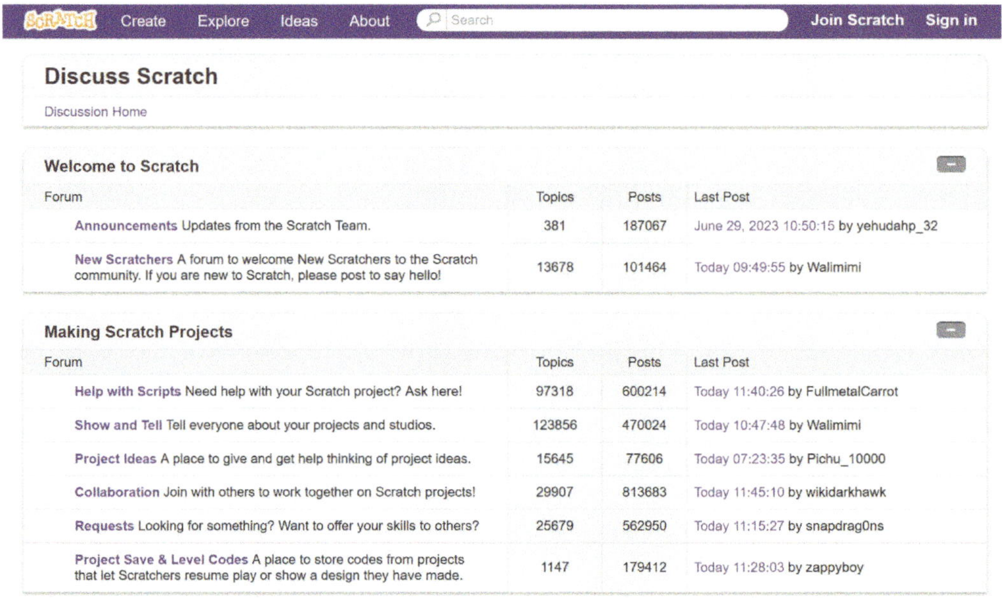

Figure 3.5 The community forum can be a great way to get some help with projects or lesson plans or to learn some new tips or tricks.

to projects of every topic, every style, every language. We can find studios or pages that curate projects to share, and we can see the most popular projects on Scratch that week.

All this content can, of course, be a distraction. Your students will undoubtedly want to try all kinds of projects they see on Scratch, and we do need to find a healthy balance of consuming entertainment of any genre on any platform. We also need to see some of this time spent as research and inspiration. Students need to fuel up their creativity, get ideas, and be both inspired and challenged by seeing other creations on Scratch. Seeing what it can do can fire up students' interest in learning to use it.

Sharing

As our students create things, they'll want to share them with friends. Because Scratch is set up to be a sharing platform, it makes it easy for them to share their creations. While this might be an opportunity for friendship and pride in their work, we can use this same system for other benefits.

This sharing system is a huge help for us as educators in following our students' work. They can submit links to let us see their projects. If you apply to get an Educator account, you can link with your students to have an even easier time keeping track of their work. Even without Educator accounts, you can follow their profiles to easily browse their projects or be alerted when

Figure 3.6 There's always lots of great inspiration to be had on the Scratch website. The Ideas page has tutorials and activity guides to help folks get started. The Explore page lets you search for either projects or studios to find something fitting a theme or concept.

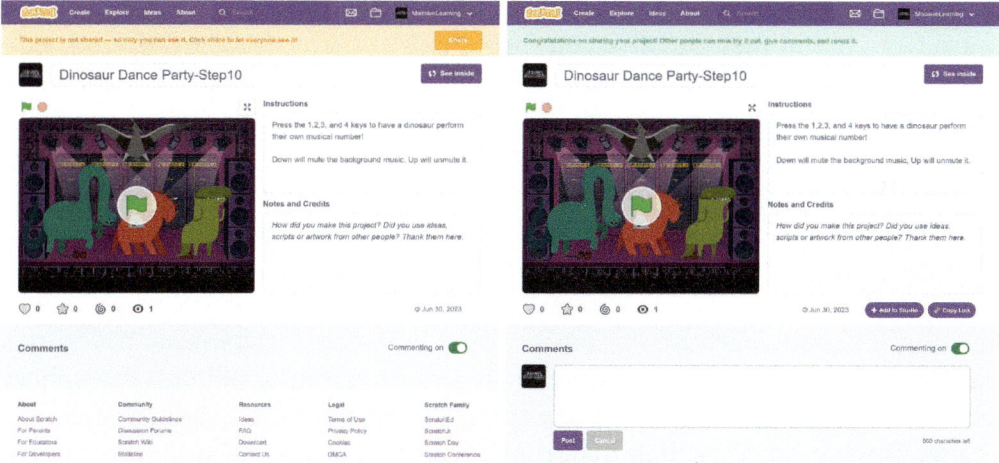

Figure 3.7 You don't have to share what you create, but it does help build the community. For Education, it can allow teachers to share starter projects or for students to do peer-review or submit projects. Here's a project before and after sharing.

they share something new. Because shared projects are still editable, you may want to create copies of your students' projects at submission so they don't intentionally or unintentionally continue working on it past submission time for your grading and review, though they could continue working on their original version. If you prefer or are using the offline version, you can also have students download save files of their projects and send you those, which you'll be able to load into Scratch to see that project as it was at the time of submission.

This system can be used the other way around as well. You can create starter projects or assignment templates and then share them with your students, who can then remix the projects to complete the assignment. This can be a great way to both inspire or test students, giving them some start to the project, and/or using comments to guide them about what you want or they should use/not use. You can send buggy projects to have them try to fix them to demonstrate understanding and skill or concept mastery.

Openness

Scratch doesn't just let users share their projects; it also lets the community see how they made projects. Every shared project page has in the upper right-hand corner a See Inside button and, if you're logged into an account, a Remix button. Both of these allow a person to open the Scratch editor with the project loaded. This way, any user can see all the code, art, and sound used to make any shared project on Scratch.

This is, of course, a double-edged sword. It means your students can look inside any project shared on Scratch and see exactly how it's coded. They can learn from it, but they can also copy from it. Most educators' minds immediately focus on the plagiarism side of this issue, but I'd argue it's more important to focus on the learning side. Yes, students can copy systems, concepts, assets, or even whole projects through this system on Scratch, but with every writing assignment, students have had access to millions of books, articles, and websites that they could copy from, too, and we still have students that managed to successfully learn English.

Scratch heads in the opposite direction for an important reason. You're reading this book because learning coding isn't easy or obvious. If you talk to any professional programmer, they'll almost certainly tell you they check others' work, ideas, and solutions through sharing and help websites like StackOverflow. Further than that, they probably learned to code from shared code; whether it was programming magazines and books that gave code examples or coding websites, they probably learned a lot of their early ideas and methods from copying examples and adapting them. Working examples are the key to understanding coding. By having a working system, early coding learners can poke it, adapt it, break it, take it apart. That's how they learn the real meaning of it.

We have to remember that the goal in coding education isn't whether the student produces a project; it's whether they understand it. Submitting a project is only part of it. Can they describe the project? Can they explain why something works or doesn't? This is exactly why we have students show their work in math; it isn't the result that matters but understanding the process. Don't let openness scare you, and make sure you look beyond the simplicity of submissions.

4

Scratch Basics

In this chapter, we'll go over a bunch of the concepts and structures in Scratch to introduce you to how it works and how to think in Scratch terms. We'll cover the basic components in Scratch, how each basically functions and how you can influence them, and clarify terms we'll be using in the rest of the book. We'll have some callbacks later in the book when we're dealing with particularly relevant applications to the principles outlined here.

Website

While there is the downloadable offline version, the main experience of Scratch is built around the https://scratch.mit.edu website. This provides a broad approach to the Scratch experience, providing an enticing landing page full of constantly changing content recommendations, a user account system, learning guidance, and a helpful user forum. The main advantage of the website over the downloadable version is the community around Scratch – you'll be able to see what people are creating, share your own creations, give and get feedback through the forums and project pages, and get helpful links to more information.

However, with students, this can present more stimulation and distraction than may be desired. You may find the downloadable Scratch program that just focuses on creation a better match for your classroom or student needs. It might be counter-intuitive, but I think it might be best to use the downloaded version for older students, to have them focus on their own creations,

DOI: 10.4324/9781003399018-4

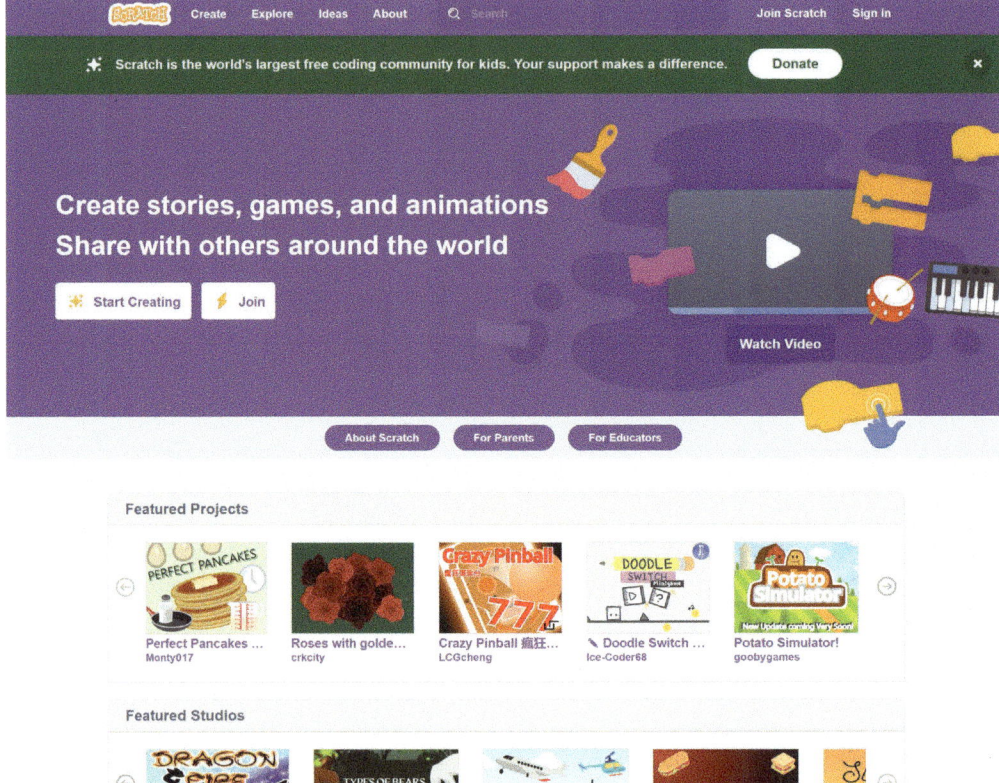

Figure 4.1 The http://scratch.mit.edu home page.

while younger students are less focused on designing their own and are more important to inspire about what they could create over the years to come.

Starting at the top left corner of the website is the Create button. This is the most important part of the front page – leading to the Scratch editor, where you can create your own projects.

Beside the Create button is the Explore button; this leads to a search page to browse through all the projects that people have created with Scratch and shared with the world. This can be a great resource for inspiration, especially when you want to deal with a certain concept or subject – whatever it is, somebody will have created something about it!

Next is the Ideas button. This is sort of a tutorial page, giving suggestions to new users about what they could try with Scratch. The Activity Guides here give an easy walk-through to some basic concepts that can be a great way to get students started with Scratch, with some common ideas that kids love. They are quick and simple, letting kids try and succeed in minutes so they can build confidence and basic knowledge.

The About button leads to a page for adults, providing deeper information about Scratch and the Scratch Foundation. You can get some insights about the team's philosophy, research, and more information specifically for parents or educators.

Below this top bar, the page will adjust depending on whether or not you are signed into a Scratch account. New, or at least not-logged-in, users will see a large banner to guide users to start creating a project, create an account, or watch an explainer video. As well, three buttons provide access to the About page, or directly to the For Parents and For Educators pages. For Parents provides a useful FAQ, some videos about Scratch, and other suitable information to get parents up to speed. For Educators provides links to educational resources, links to communities of practice (including a Facebook group and ScratchEd meet-ups), news, and a link for creating a teacher account.

If signed in, this section instead will provide news about your account. It will list recent activity by other accounts you follow as well as list Scratch News to keep you up-to-date about Scratch events like Scratch Week.

Under this head banner we have the real draw of the Scratch front page – the constantly changing recommendations! The Featured Projects, Featured Studios, Scratch Design Studio, What the Community Is Remixing, and What the Community Is Loving sections provide a look into what people around the world are creating with Scratch. This can be a wonderful inspiration and insight into what Scratch can do. This can be a really heart-warming place to check out and see what ideas kids have brought to life with art, music, and code, sharing stories or glimpses of their cultures. As well as a fun place where some of the most polished projects created in Scratch come to shine, whether they are games, animations, or musical experiences. Of course, this can be a bit of a distraction for some students, but it can also be a useful inspiration. Something you can have students explore during indoor recesses, or at home, at the very least, and something you can draw from for inspiration.

Accounts and Saving

If you have a free Scratch account and are using the online Scratch website, your project will autosave every so often to your online account. Online accounts are a great way to help students, or yourself, save and share projects. Without an account, you can still save your projects but will need to manually download a save file and manually upload it if you want to work on or run it again. With the downloaded offline version of Scratch, you'll save or load files of your work just like any other downloaded software.

Project Pages

With online accounts, projects are given project pages. Project pages help present the project to other users and allow a creator to provide some context. Besides users being allowed to run the project, there are Instruction text fields as well as a Notes & Credits text field to provide more information about the project. Users can also like, favourite, and remix projects on the project page. For a project to be visible to other users, a creator must choose to share their project.

Remixing

Scratch users can remix a project from its project page. Remixing makes a copy of a project that a user can then modify as they see fit. Remixes will link back to the project they were remixed from (though there are ways around that, so don't count on it for stopping plagiarism). Remixing might seem like a detriment for education (as a built-in system for copying), but it has a lot of advantages too. With every project on Scratch being openly shared

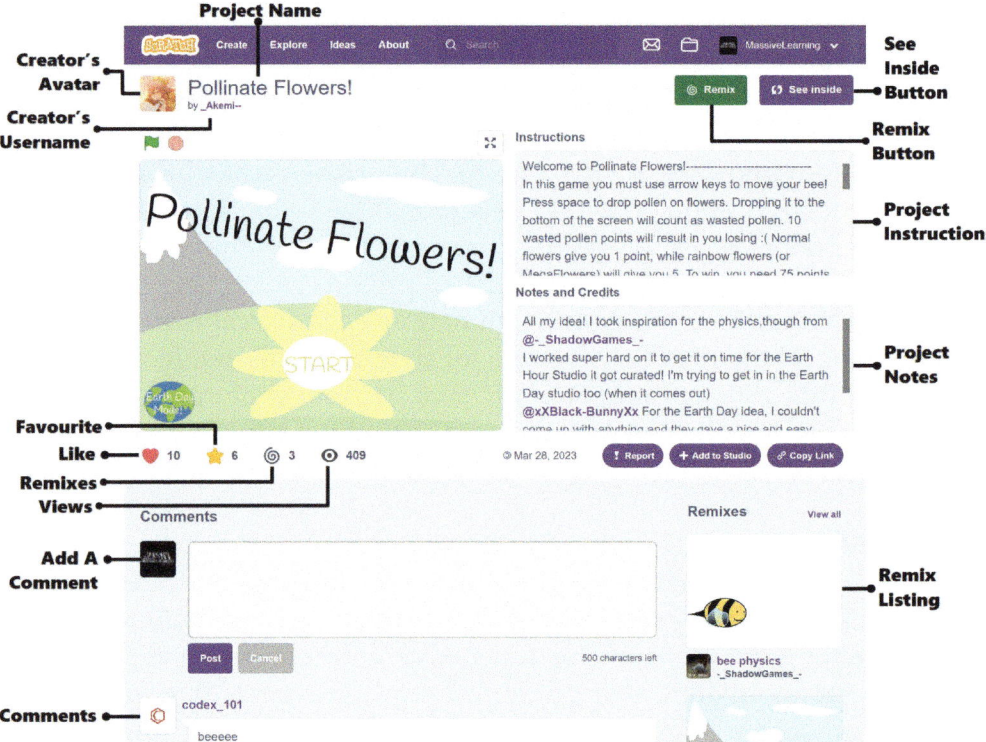

Figure 4.2 A publicly shared project page. Every shared project can be played, explored, and even copied and altered by other users. You can like, favourite, remix, or add comments to public projects.

and observed, students (and teachers) have millions of creators that they can look to for inspiration and advice. We can look through the code to try to understand any project we like or have questions about. Learning to read and understand others' code, or even adapt or correct it, are very important skills to develop, so this open-sharing system offers a nearly endless opportunity to practice these high-level skills. While we can't rely on the exact code turned in as a flawless metric of effort (when sharing is so easy), we can turn our attention to the more important question of whether the student understands the code. We can see if they can express the intent, process, and outcome of the code in their projects as a measure of true understanding.

The Scratch Editor

The main feature of the Scratch website, and the sole purpose of the offline Scratch program, is the Scratch editor. This is where a Scratch user can create their own projects. It contains the tools for managing what components or objects are in a project, a view of the project, tools to code interactivity for the project, art tools to create visual assets, sound tools to manage and edit sound assets, and the ability to add other tools or extensions for added functionality. Needless to say, this is a lot to cover and learn, and understanding and using it is the focus of 99% of this book series. Master the editor and you'll be able to create whatever you can imagine, giving you an interactive platform for

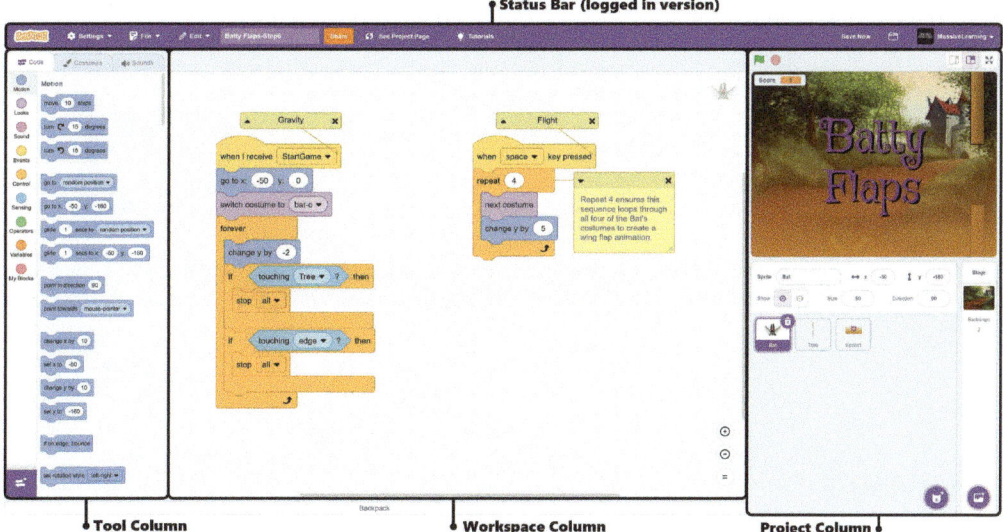

Figure 4.3 The main sections of the Scratch editor.

deep integrated learning that you can use for any subject. There are multiple components to the Scratch editor – the Status Bar, the Tabs, the Stage Window, the Properties panel, the Backpack, the Sprite Listing, and the Stage. We'll dig deeper into some of these in their own sections, which follow, but let's take a look at a couple of the more simple components here – the Status Bar, the Tabs, and the Stage Window.

The Status Bar

At the very top of the screen is the blue Status Bar. Like most programs, this top bar allows you to access some core functions about how the editor works, or what you're working on. At the left is the Scratch logo, which will take you back to the home page. Then the Globe icon allows you to switch languages to any of the 74 languages Scratch has been translated to. The File option allows you to start a new project, automatically saving the project you're working on to your account, if applicable, or to load or save projects, either to your hard drive or optionally to your account if you are logged in. This is followed by the Edit button, which can allow you to undo some actions or to turn on or off Turbo Mode. Turbo Mode makes projects run at super high speed unrestricted

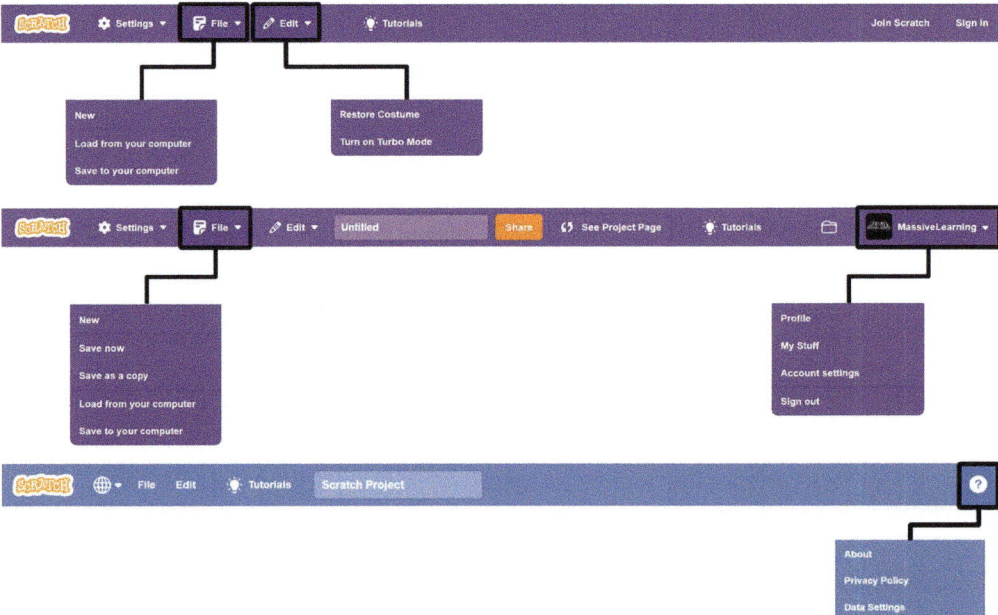

Figure 4.4 Three versions of the Status Bar. The guest version at the top doesn't allow any cloud saving, loading, or sharing, or any access to accounts. The middle is the version for logged in users, where they can save or load from their accounts on the cloud, as well as share their projects. At the bottom, the offline version with no cloud, account, or sharing options but includes options for About, Privacy Policy, and Data Settings for the offline version.

by Scratch's usual frame rate, which can make some amusing things happen. A Tutorial button will take the user to some recommended starter projects in case they're confused or overwhelmed by the potential of the editor and the blank screen starting point. At the right side, one can Join Scratch or Sign In if one is a "guest", that is to say, someone not logged into an account.

If the user is logged in, a number of minor changes occur. The File button will provide account-based saving and loading of projects to the cloud. After the Tutorial button, there will be a text field to name a project, with "Untitled" as the default. This will be followed by the Share (or Unshare if already shared) button, which will make the project available (or unavailable) to the public. The See Project Page button will allow the user to preview and edit the project page for the project for when it is shared. This can allow the creator to provide notes to future users, such as instructions on how to play a game, or provide credits for the project. When logged in, the right-side options change to Save Now; a Folder icon, which takes the user to their account's My Stuff or project listing page; and finally, their icon and username as a button. When clicked, they will get the option to go to their account profile page, their My Stuff (or project listing page), their account settings, or to sign out of their account.

Design Tabs

The Scratch editor allows us to create something as complex as interactive media because of the breadth of its abilities. With it we can create and edit code, art, and sound to bring our projects to life. We can do all these wide-ranging tasks thanks to the tabs in Scratch: Code to influence objects' behaviours, Costumes or Backdrops to change an object's appearance, or Sounds to work with the audio the object can play. Each tab, located at the top left corner of the editor, allows us to switch our focus to a different multimedia aspect. Though most of our focus is on the Code tab, we will incorporate some sound effects through the Sounds tab, and we'll need some custom art and teach a little bit about the Costumes/Backdrops tab and its art tools. Learning to work in all three tabs will help you unlock all the potential of Scratch. We will cover more about each tab in depth later in the chapter.

Stage Window

Projects either on the project page or inside the editor are displayed in an area we're going to call the Stage Window. This window acts like the video canvas on a YouTube page, or the Player area on a HTML5 game page like Kongregate. All content and interaction happen in this little 480-pixel-wide-by-360-pixel-tall area. All Scratch projects are the same size, so just like when recording a video or taking a picture, you may need to think about what you

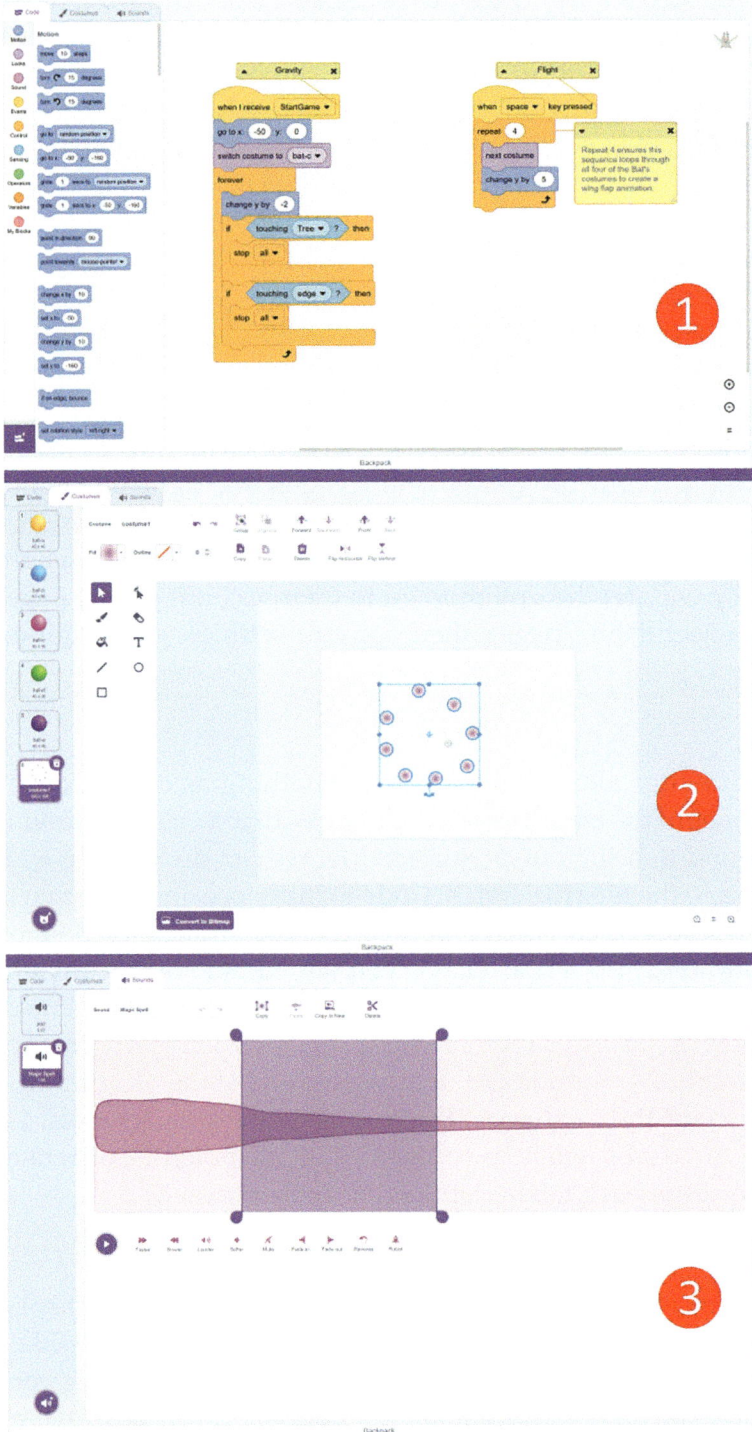

Figure 4.5 A comparison of the three tabs in the Scratch editor: ❶ Code, ❷ Costumes, and ❸ Sounds.

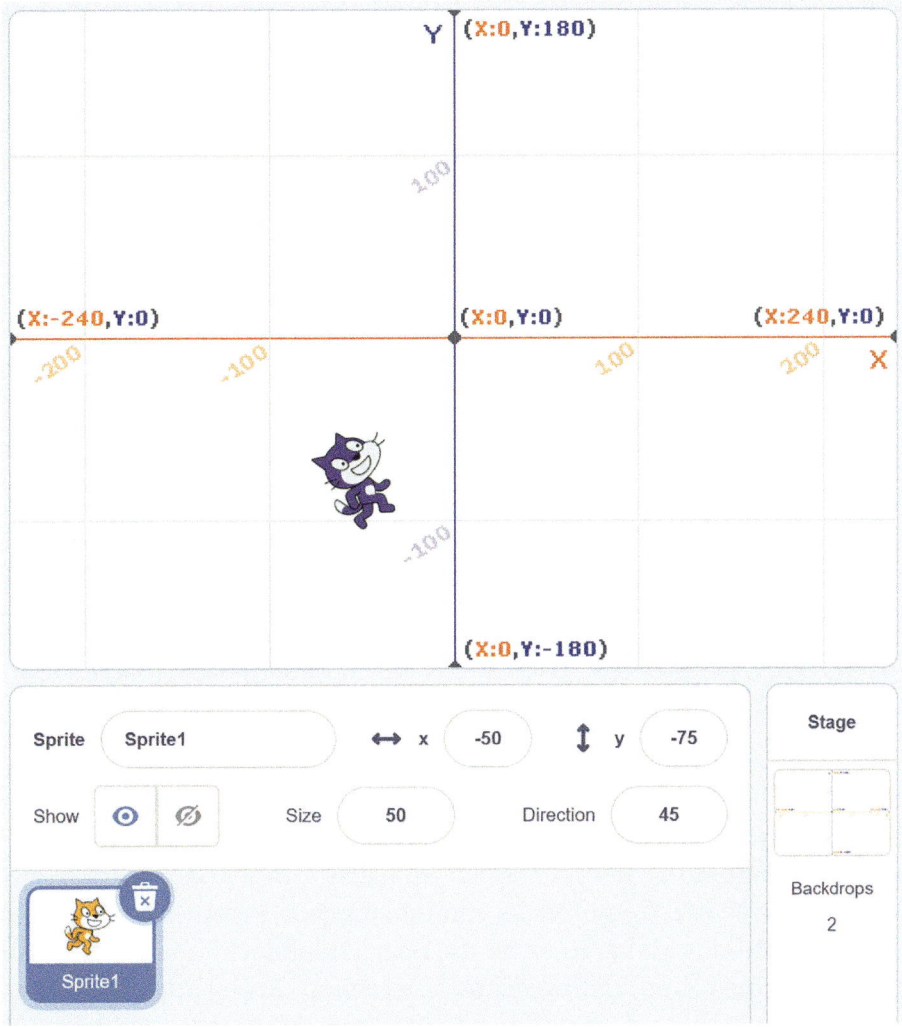

Figure 4.6 A look at the Stage Window with the XY grid backdrop selected showing the four-quadrant cartesian plane used for positioning objects in Scratch.

want to show and how to fit it in the view. You can expand the view, but it just stretches the content to a visually larger area. While working on a project, you position all its components within the Stage Window and can test it by running it in the Stage Window to see how it works, tweak things, and play-test. Above the Stage Window you'll see a green flag (▷) and a red octagon or stop sign; these function to begin (play) the project or stop it.

Everything in the project has a grid coordinate position consisting of a left/right X-coordinate and an up/down Y-coordinate. The Stage Window is split into four quadrants, with X:0 Y:0 in the centre. X is negative to the left of centre, positive to the right. Y is positive above the middle and negative below it. The grid extends infinitely for any objects that go off-screen, so any object can always be tracked on this grid, but Scratch tries to keep objects visible on the screen and not go past the edge. Each point on the grid is equal to 1 pixel. When a sprite moves 1 step, it moves 1 pixel on the screen and its coordinates change by 1. One of the built-in backdrops for Scratch is a diagram of the grid that can help visualize this system.

The Code Tab

The Code tab is where we can add and edit code for Sprites. This is how we achieve interactivity, animation, or any other activity in the project. Code is the key to bringing any project to life. Even assets added through other tabs must be called for or implemented through code. This tab has the most complexity and depth to its capabilities, but this book is all about explaining these features, methods, and practices, so for now we'll just explain how to navigate the editor, and the projects will help showcase how those features and abilities come to life. The way you'll use the Code tab is through block coding, which is a form of computer programming done through code blocks.

Code blocks are the heart of Scratch, a brilliant idea to make coding more accessible, and especially kid-friendly. We use them to build our project's interactive components, taking Scratch projects from digital art and sound to digital interactive art. Code blocks can be snapped together or pulled apart like plastic building bricks. To assist the user, code blocks shapes and colours help provide clues about their use and interaction. Each code block provides a simple instruction or data point, and sticking the blocks together, we can craft a set of instructions for the computer. We use code blocks to determine when things should happen, as their interconnection gives the computer a sequential order of operations to follow, while we input data points to customize the instructions to our exact needs.

Code Block Library

When you've selected the Code tab along the left side of the screen, you'll see the Code Block Library. You can scroll up and down this section to see all the available code blocks for the sprite or stage, depending on which you've selected. To the left of the code blocks is a category shortcut system. All the code blocks are colour-coded into categories based on function. You can click on the coloured dots to jump to a specific category rather than scrolling through the list. Click and drag a code block from the Code Block Library to

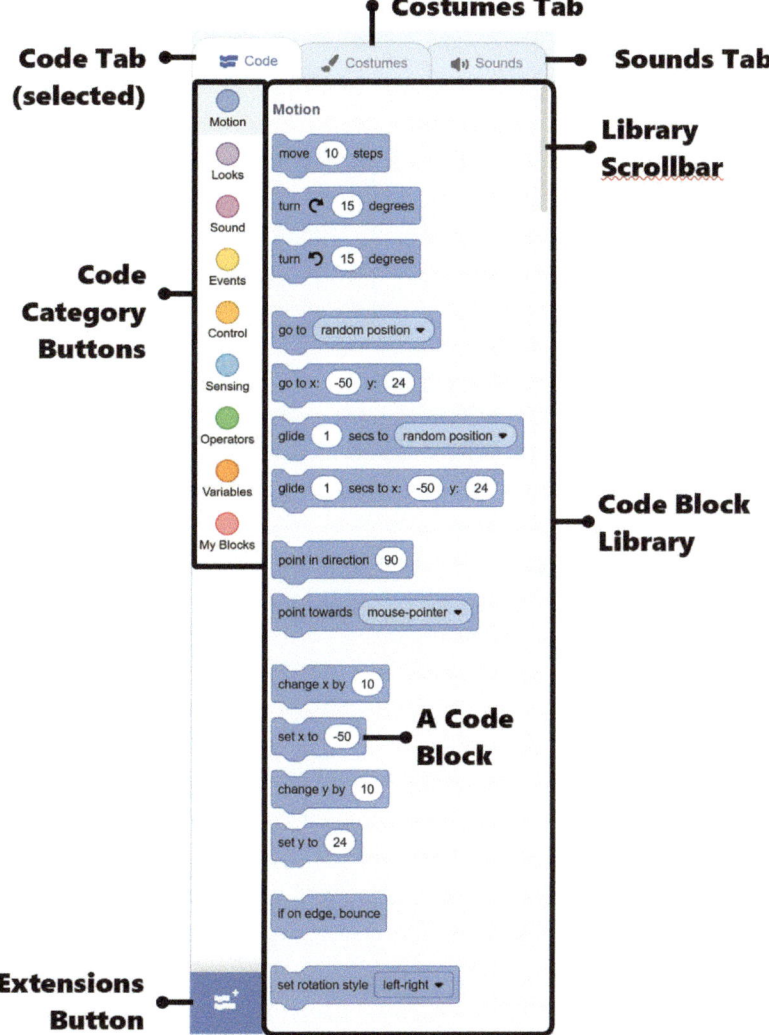

Figure 4.7 The left side of the editor when the Code tab is selected, showing the Code Categories shortcuts and scrollable Code Library.

the coding workspace to use it. Code blocks can be dragged from the workspace back to the Code Block Library to delete them from your project.

Coding Workspace

The large empty area in the centre of the editor, at least when the Code tab is selected, is the coding workspace. The coding workspace, as you've probably guessed, is where you assemble all your codes. As you click and drag code blocks out and connect them, you'll build your instructions for the computer on how your project should work.

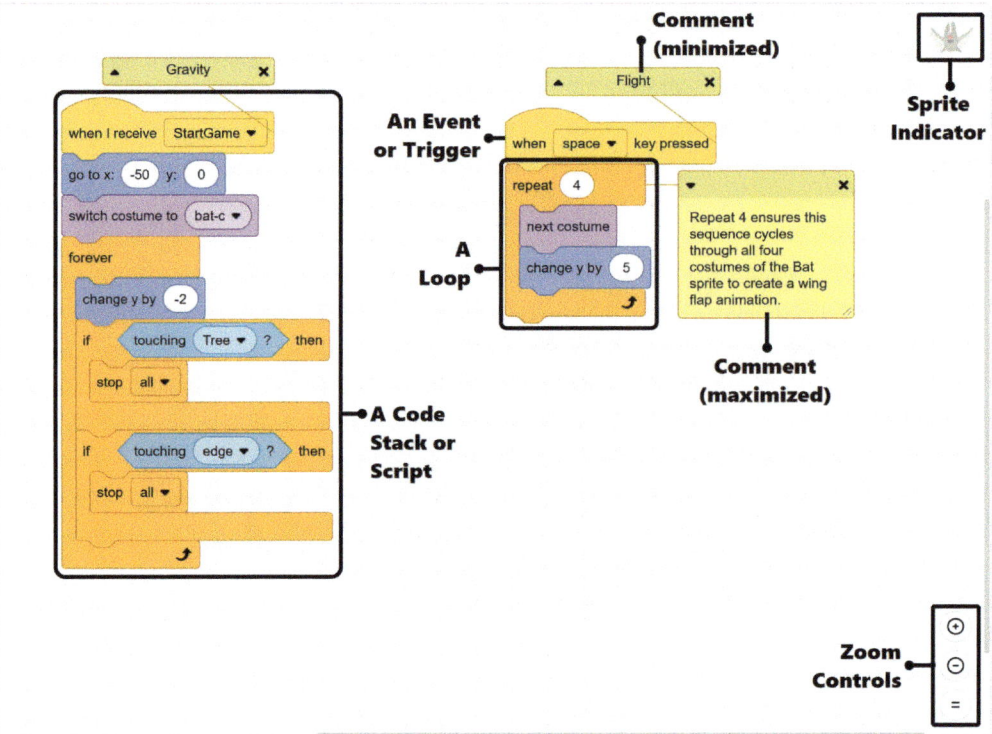

Figure 4.8 The Code tab's workspace, showing some example code stacks or scripts, as well as some comments.

All your code will be explorable through the coding workspace. You can click the Zoom controls at the bottom right-hand corner of the coding work-space to zoom in (+), out (-), or reset the zoom (=). To scroll around the area if you need more space, you can click (and hold) on any empty area to drag the view, use the scroll bars along the right or bottom of the coding workspace, or use your mouse wheel to scroll up and down. This means having all the space you need to work, expanding in any direction. Just be careful, though; more than once I've been stumped by a bug in a student's project, only to find after an exhaustive search that they had some code hidden way off to the side that they forgot about.

Lastly, it's also important to remember that code is assigned to specific sprites. The code you see in the coding workspace is for a specific sprite, or the stage. You can tell what object is selected by the blue outline added to either the stage area of the editor or to a specific sprite thumbnail in the Sprite Listing. In addition, Scratch gives you a little watermark or ghost image of the selected object in the upper right-hand corner of the coding workspace as a reminder. The first time your students add a new sprite to a project, they may be disconcerted that all their code has disappeared! Of course, this is just their view changing; the code hasn't been lost at all, as they just changed what object was selected. Switching objects will show them the appropriate code.

Colour and Categories

The code blocks are all colour-categorized to make things more visually distinct and appealing. This helps users in a few ways. Firstly, it enables organizing the hundred different code blocks in some sensible system. Secondly, we can better distinguish different code blocks (though colour-blind students may have some issues with some categories). Thirdly, it makes it possible to quickly scan code for relevant information or compare code to find differences. Lastly, it makes it more visually interesting and, in doing so, helps reinforce categorical concepts and code structures in learners. When you get to know the different categories, you'll be able to easily jump around and find the code blocks you need for given tasks.

◆ •**Motion.** Deals with moving and positioning sprites (their direction, X position and Y position properties). Not available for the stage.

◆ •**Looks.** Deals with how a sprite or the stage looks (change the costume or backdrop assigned to the sprite or stage, respectively), displays text to the user, changes the size property of a sprite, applies a special graphical effect, change the visibility or layer of a sprite.

◆ •**Sound.** Plays or stops sound effects or music (sounds assigned to the sprite/stage), applies an audio effect to the sounds played by the sprite or stage, or affects the volume of the sounds played by the sprite or stage.

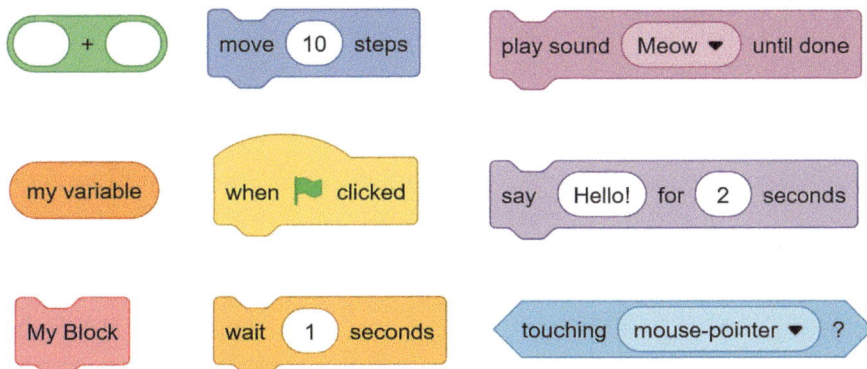

Figure 4.9 Navigating code blocks is faster and easier clicking on the categories, but you can also scroll through all the code blocks with the scroll bar. It also helps to call out categories to help students find code blocks faster when learning.

- ◆ •**Events.** Controls when things happen when code is run. Reacts to user input or other conditions in order to trigger running code. Every script (or stack of code blocks) must start with an event to know when to run.
- ◆ •**Control.** Adds complexities to the order of operations, like delaying, looping, or conditionally running code, stopping code from running, or using clones.
- ◆ •**Sensing.** Allows testing conditions within the project to use with •Controls, gets data to work with in the project, or alters how the user can provide input to the project.
- ◆ •**Operators.** Provides mathematical or logical operations for data or condition tests.
- ◆ •**Variables.** Allows the creation of variables to track data in the project and provides code blocks to use, change, show, or hide the data.
- ◆ •**My Blocks.** Allows the user to create custom functions they can then easily reuse in their projects.

Shapes and Connections

The shape of code blocks determines how they can fit together. This helps new users understand what they need and how to structure their code. All similarly shaped code blocks have to be functionally interchangeable, which means the shapes end up representing fundamental coding concepts. New users can simply follow the shapes to know how code blocks will fit together and, through practice, come to expect the pairings, learning the underlying structures of code without even realizing what they're learning. Coding logic and structures simply become natural through this almost-tactile visual system. Most of Scratch's users probably never think about the exact nature of the code blocks, but as educators, we can take the time to understand the underlying principles involved. There are five basic shapes of code blocks, with some minor variations to them.

Hat Blocks. The hat blocks are code blocks that have a bump or smoothed top, so no code blocks can attach above them, but have a bump at the bottom to connect to code blocks below. Most of the •Events code blocks are hat blocks, the •[When I Start As a Clone] and •[Define [MyBlock]] code blocks are also hat blocks. These are triggers, things that initiate the running of the code attached to them. They give a definition to the computer of a time, event, or condition that should trigger running some code. All scripts (stacks of code blocks) must start with a hat block in order to run.

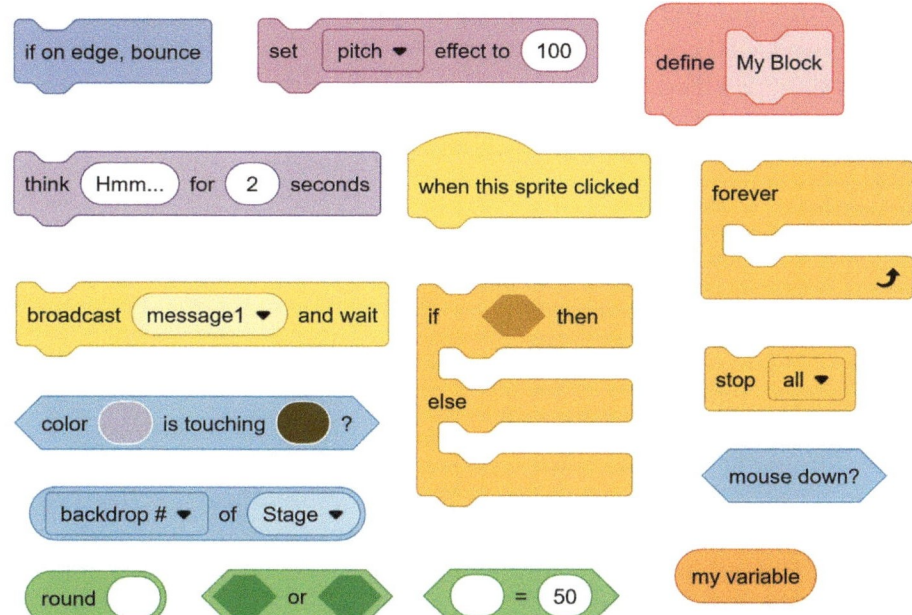

Figure 4.10 An example of some of the shapes of code blocks, including stack blocks, cap blocks, c-blocks, Booleans, and pills. The shapes tell you how code blocks can or can't fit together, sometimes even in ways you don't expect!

C Blocks. C block code blocks have a "C" or "E" shape that can contain, or nest, code inside them. C blocks are control structures, having some influence on the way the code inside them runs. They can loop it over again or run it only on a condition. These containers make it obvious for new programmers how control structures nest code, creating a distinct container for it, surrounding it to visualize that containment and control they have over it. The ●[Forever] code block has a unique distinction of having no bottom bump; as an infinite loop, it is never finished, so no code can follow it. The ●[If <condition> Then {} Else {}] code block has two containment areas because it has two possible outcomes, the condition being true or false (also known as Boolean): the first runs only if the condition is true, the second only if the condition is false.

Stack Blocks. Stack blocks are the most common code block. These are the flat rectangular blocks with a dent on top and bump on the bottom, allowing them to fit into coding sequences anywhere below a hat block. They are independent functions with no particular restriction on use, though some combinations may not produce visible results on their own. Many stack blocks have at least one space inside it for an input, a piece of data that determines some aspect of how the stack block's functionality will be implemented, setting either a target for it or a scale of magnitude. The •[Stop [All]] code block is unique in that it has no bump on the bottom, as it stops all code from running, so no code can follow it.

Boolean Blocks. These sharp, spikey-ended code blocks are also known as Booleans, or true-or-false states. They fit inside some of the C blocks, as well as the •[Wait Until <>] stack block, and could be used in some •My Blocks. They are often used to create conditions, creating a test statement that will be checked by the computer to see whether at the time of it running is found to be either true or false. This true-or-false outcome can then be used to power a conditional C block. If something is true, the •Control/•My Block will allow code to run in some manner; if the condition is found to be false, it either won't run or will run a false outcome sequence of code. Boolean blocks combine with •Controls to allow us to create dynamic and adaptive programming, by creating reactive programming through conditional tests that change how code is run.

Reporter Blocks. The round-edged blocks are known as values, inputs, or reporters. They represent data in Scratch. Every white round-edged space in code blocks can fit them in or replace them with a hand-typed data point. These are how we adjust the scale, magnitude, or other required data points in other code blocks. Some reporter code blocks aren't values in themselves (like almost all the •Operators code blocks), but rather, they are ways of adjusting a value, such as a mathematical operation.

When we connect code blocks, they become a "script" or "stack". A *script* is a sequential series of instructions for the computer. No code block works alone. It's either a trigger, like an event that detects when it should run, or it's a functional block that does something, but without a trigger it won't run on its own. Combining code blocks by snapping them together makes them functional. A script needs to start with a trigger so that it will run, but then it needs whatever code blocks will actually do the work that's needed.

When code blocks are moved near each other, Scratch will give some kind of guide about how it will join the code blocks together if they are released at that position. Code blocks that connect above or below will show a grey shadow in the position they will snap to. Code blocks that connect inside a code block like a Boolean or reporter will show the space they will snap into outlined in white. When you're trying to snap code blocks into specific places, like when you're using an •<<> **and** <>> block, you can look to the white outline to make sure it's going to connect correctly before you let go.

You will, at some point, need to pull apart code blocks that are connected. If you click on a code block and drag it, all code blocks attached underneath it will move with it. This can be handy for rearranging your scripts, but it can be annoying when you just want to remove one code block out of a script. You'll need to get used to dragging out a stack of code blocks then pulling off the bottom any code blocks you didn't want to remove, then putting them back. You can also right-click on a code block and choose "Delete Block" to remove it from the middle of a stack. This will delete that specific code block and any internally connected blocks, but not the blocks that follow under it, and will instead move those up to keep the script contiguous.

Values and Data Input

Many code blocks have customizable portions where the creator can enter a number, text, event, or object that provides a key detail to the code block's operation. These values or arguments provide important data that allow a generic function, such as Move to know how far to move, a Say block to know what to say, and so on.

There are five basic types of data points used in code blocks.

Reporter. Our round-edged blocks fit into all the white round ovals in a multitude of code blocks. When a number is needed, white oval spaces can be filled with either typed-in values or filled with a reporter code block. This includes text values, also known as "strings", that can be typed in text to display, or have text saved in variables and use a variable reporter code block to insert that variable data into a code block. The •[Glide (#) secs to X: (#) Y: (#)] code block has three different values, one for the time duration, the other two for coordinate positions, all that can be specified through a typed value or by a reporter code block.

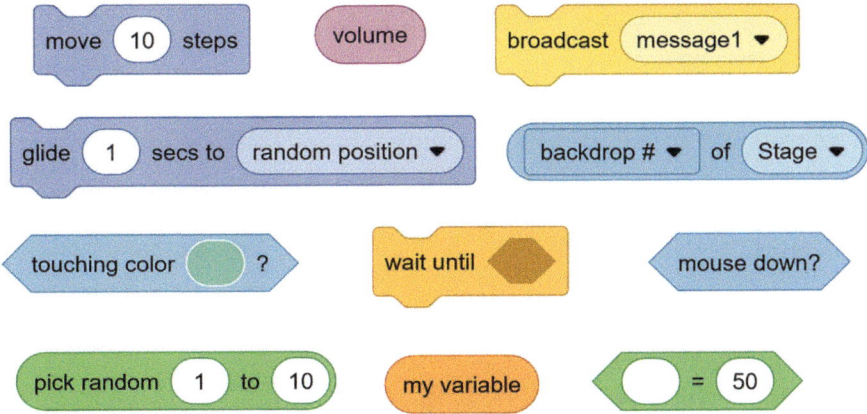

Figure 4.11 A sampling of how inputs are used in Scratch. Code blocks can have controls or inputs to help define their function, or code blocks can be the input for other code blocks!

Dropdown Lists. Some code blocks need values, but the options are restricted to a defined list. These dropdown values allow the creator to select the appropriate option. In the ●[Set Rotation Style (*style*)] code block, there are only set methods of determining rotation style, so the user must choose one of the three options. In the ●[Start Sound (*sound*)] code block, any sound associated with the sprite can be selected, and this list updates any time a sound is added or removed from the sprite. The ●[Broadcast (*Message*)] code block lists the messages already created in the project but also allows it to select a "new message", which brings up a pop-up so they can be named and create new messages that will be added to the list. So keep in mind, some lists are always the same, and some dynamically adapt to the project.

Reporter/Dropdown Lists. Some code blocks have a combination Reporter/Dropdown List. These allow the user to select a preset option like a dropdown list but can also accept a reporter-shaped code block to dynamically, or programmatically, adapt. A good example is the ●[Switch Costume to (*costume*)] block. Here the creator can select from the dropdown one of the costumes for the sprite, or they can use a reporter code block to give a number that will select the costume by its number. This can, for example, allow a ●(Pick Random (#) to (#)) to be used for a random costume, a variable, or a mathematical calculation determine the costume.

Booleans. A number of ●Control code blocks change the way code runs on the basis of evaluating a condition. A Boolean block–shaped hole, like in the ●[Wait Until <condition>] code block, allows users to insert a conditional statement. These conditions are always true or false (or "Boolean") and are made from operator or sensing code blocks. These can be used to create comparison or test conditions that are calculated as true or false when the code runs. As an example, the code block ●<touching (*mouse-pointer*)> will determine if the sprite that's being coded occupies the same screen space as the tip of the mouse cursor. Often, one will need to combine multiple code blocks using some of the numeric comparison blocks in the operators, like ●<(0) = (0)> to compare two values, such as ●[If <●(Player Lives) > (0)> Then] to test a particular condition relevant to the project.

Colours. Lastly, some code blocks use colours as referenced values, where a specific colour is used to determine something. The ●<colour (colour) is touching (colour)?> code block is a conditional test that checks every pixel of the sprite being coded for the first colour and determines if any of those pixels are in the same position as a pixel of the second colour (whether it's from another sprite or the backdrop). This allows some very precise collision detection between objects so you could, for example, tell when the grey tip of an arrow hits the red bullseye of a target even though the two objects overlap on lots of other pixels. The colour values are always set by clicking on them to bring up the colour panel, where a user can either select a colour or use the colour picker tool to get the exact one from the Stage Window.

Events, Triggering, and Timing

Code blocks aren't just about what they do; they need to do it at the right time and in the right order. Paying attention to the flow of code is important to have things happen correctly and reliably. All codes, in order to run, must have a trigger that provides instructions to the computer on when it should be implemented. This is often done through the ●Event category of code blocks but can also happen through the ●[When I Start As A Clone] ●Control Block, or the ●My Blocks. We use triggers to call a script that can be from direct user input, like ●[When ▷ Clicked] or ●[When [Space] Key Pressed], or something more incidental, like ●[When (loudness) > (10)]. Lastly, they can also be from direct commands to run other scripts, like the ●[Broadcast [Message]] or ●My Block code blocks.

Triggers are the most important function of determining the flow of code, but we also have other code blocks and methods to changing the simple sequential flow of code through a script. The ●Control code blocks are all about affecting how code flows. They contain ways of delaying, repeating, stopping, copying, or conditionally running code. By using these code blocks, we can create much more dynamic, reactive, and complex flows to our projects. Mastering these techniques can allow you to create tremendously powerful and deep projects.

The Costumes or Backdrop Tab

The Costumes tab, which renames to Backdrop tab if the stage is selected, is a wonderfully rich tool for creation. Here, the user can create and edit digital art assets to use in Scratch, either costumes for sprites or backdrops for the stage. With the Costumes tab as a built-in, fully functional digital art tool, there's a lot to learn and try with it, and I've added some practice and training exercises in a number of our projects in this book. I'm about to give a short rundown of the features, components, and methods involved in creating your own digital art in Scratch. Whether you're creating costumes for sprites or backdrops for the stage, all of the following apply equally; we'll just shorthand everything here to Costumes/Costumes tab for simplicity.

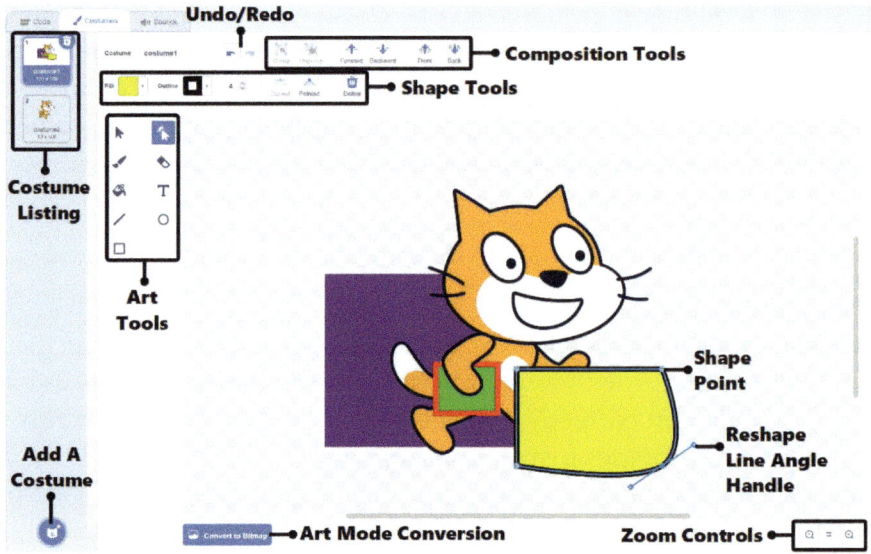

Figure 4.12 The Costumes tab workspace, with an example of using the Reshape tool.

Asset List

When you switch to the Costumes tab, you'll see all the costumes attached to a sprite listed on the left-hand side of the screen, called Costumes List. Depending on how many are there, a vertical scroll bar will let you go through all the costumes. Each costume has both a Costume Name and a Costume Number property. The Costume Name, you can set in the Name field in the upper left-hand corner of the art workspace. The Costume Number is assigned based on the order of the costumes. You can change the order by clicking on and dragging costumes up and down. In the code, you can refer to costumes either by Costume Number or by Costume Name. Costume Number being a number that can be mathematically adjusted, you can, for example, ●**(Pick Random (#) to (#))** to select a random costume. For human readability, though, Costume Names are generally easier to work with. The order of costumes will matter if you use the ●**[Next Costume]** code block, which will simply switch to the next costume in the Costume Listing.

Sizing and Centring

The canvas is where all the art is created. A space to place whatever shapes and colours comprise the costume or backdrop. The canvas provides a large area for creation and then an inner rectangle that defines the actual Stage Window size and shape. You can use this as a guide for sizing your artistic creations. If you are making a backdrop, you'll want to completely fill the Stage Window size guidelines. If you're making a sprite, you can approximate how much of the screen the sprite will take up at 100% size. Through the Properties panel or in code, we can adjust the size of a sprite, but the default scale of a sprite can be useful to know as well.

In the centre of the canvas is a small (+) reticle or target. This marks the centre point of the costume, and if the sprite rotates, it will rotate centred on this spot. It is also the exact spot that the sprite's position is based on. For rotating sprites, you'll likely want to centre them. For games in a side-view perspective, you may want to place sprites so they stand directly on the reticle.

Scaling and Rotating

If you select an object or group of pixels, it will be outlined in a blue box. At the cardinal and corner points, you'll see dots on this box. You can click and drag these points to scale the selection. The cardinal points allow you to scale it in only dimension: either height or width. The corners allow you to dynamically scale it in both height and width while maintaining the aspect ratio (or width:height). Aspect ratios are important for keeping an image looking correct despite re-scaling it. If you change its width differently from the height,

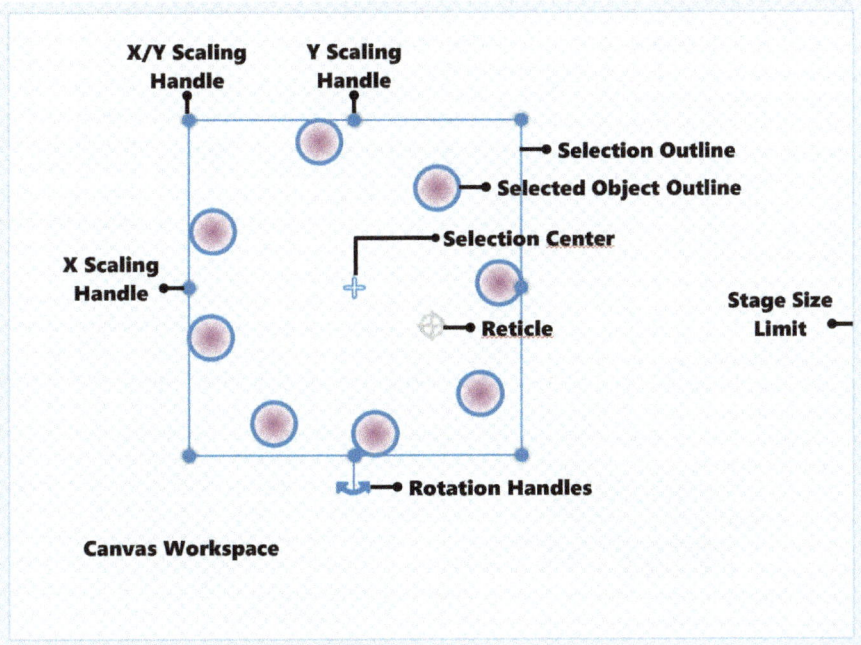

Figure 4.13 An example of working with selections in the art canvas. Note the selection's centre and the canvas's centre are marked differently. Often, you want these to be aligned on top of each other for accurate position and rotation.

it may look stretched wide or squeezed narrow. To dynamically change the scale ignoring the aspect ratio, you can click and drag a corner dot while holding the Shift key to allow free resizing. By holding Alt while resizing, you can scale out from its centre rather than a single side.

Keep in mind that while selected, at the bottom of the blue selection box, you'll notice two small curved arrows: the rotation tool. Click on this and then drag to the right or left to rotate a selection. In bitmap mode, this will allow you to rotate bitmap pixel selections, but it may create jagged lines.

Vector vs. Pixel

There are two kinds of digital art in Scratch: vector art and pixel art. You will see under the canvas a button to switch modes, depending on what the current art is. If you are working on pixel art, a Convert to Vector button will show. If you are working on vector art, a Convert to Bitmap button will show. Both art forms have their unique place, work in different ways, and depending on your project, you may want one or the other to best suit your needs.

Vector art is made through vectors, mathematically defined points, lines, and areas. It works by defining the shapes and colours of art mathematically using relative units, allowing it to be scaled to any size without any change in the quality of the art. It also means that shapes are distinct objects, collections of mathematical rules that define it. Vector art can be edited by changing the points or definitions of lines or colours. It places the focus of creation on the shapes that comprise an image. The distinct object nature of shapes in vector art allows each individual component of an image to be edited separately quite easily. Almost all professional graphic design is done with vector art, so it can be a very useful skill for students to learn.

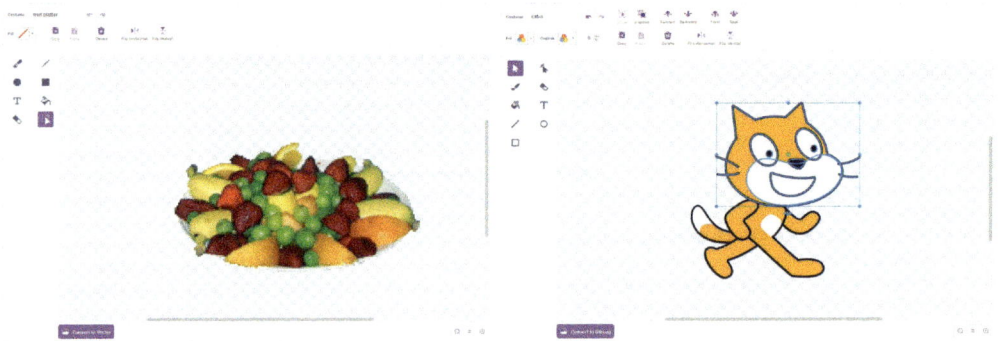

Figure 4.14 A comparison of the two art modes – pixel on the left, vector on the right. Notice how the art and composition tools change with the mode. Kids will often find pixel easier to work with, but vector offers some very handy tools and methods, so try to encourage exploring both!

Bitmap, or pixel, art is made up of a grid of single-colour dots, or pixels. Pixel art is famous from early video games, with simple or blocky art, but it can be a higher definition in Scratch than the early video game days. Pixel art allows creators to paint freely with colour, applying it to the grid without any separation of objects. Creators can simply paint over any existing pixel to change it, and the image always stays just one single layer of paint, never becoming more complex or harder to edit. On the other hand, with every pixel being treated equally and nothing separated into shapes, it can make some editing harder (and you don't get the Reshape tool when working in pixel art) but can make other changes easier. Often, students will prefer the simplicity of working in pixel art, as it has a lot of similarities to drawing by hand on a piece of paper, but they should be encouraged to try to learn the vector tools so they can understand and use both.

Art Tools

The art tools are the key to making your own art in Scratch. They'll offer you power to unleash your creativity. Here I'll give a brief explanation of how to use each one and how they function in each art mode – vector or bitmap.

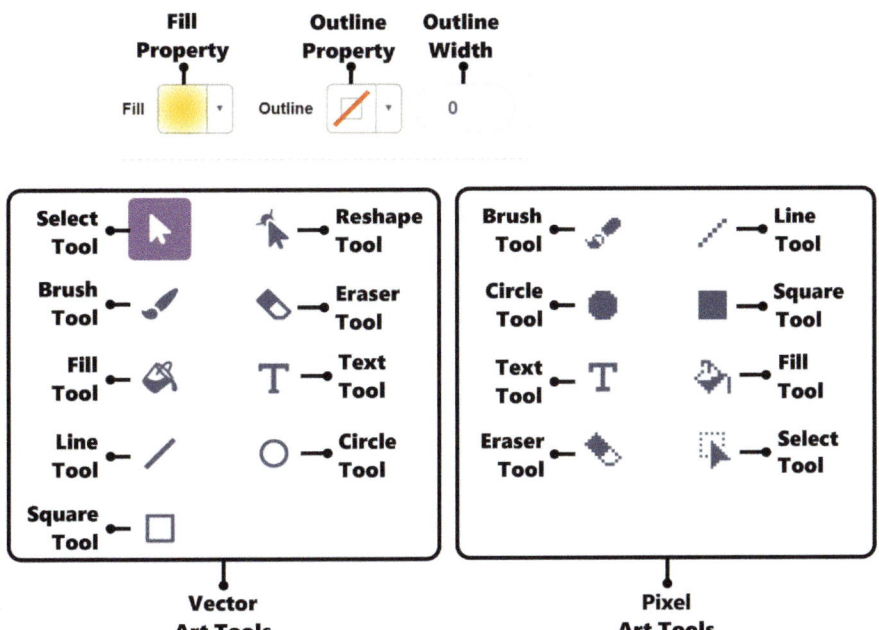

Figure 4.15 The art tools for Scratch. Shape tools define the colour and style of a shapes colouring. The art tools available will depend on the art mode selected, either vector or pixel.

Select Tool. The Select tool allows you to choose a shape or other object in vector mode so you can alter its properties or move it around the canvas by dragging it or using the arrow keys. Selecting an object is done by simply clicking on it. When selected, a box is drawn around the object with dots at the cardinal and corner points that you can then click and drag to resize, or use the Rotation tool at the bottom of the selection to rotate it. Also, when it is selected, an object's fill, outline, and layer can be changed through the appropriate buttons in the editor. Multiple objects can be selected by holding the CTRL key while clicking on multiple objects, or clicking and dragging to form a selection box, where all objects it touches will be selected. When in bitmap mode, the select tool always forms a selection box and allows you to select a group of pixels which can then be moved, resized, rotated, or deleted.

Brush Tool. The Brush tool allows you to freely draw on the canvas. What you draw is based on the Fill Colour and the Brush Size properties. Select a brush size to change how big or small the brush is, which creates a circle of colour that many pixels across. The Fill property determines the colouring used. While drawing only a single colour can be selected. But after a shape is drawn, the shape can be selected and have its fill colour set to either any single shade of colour or a gradient of two colours, either horizontal, vertical, or radial in pattern. The brush creates a complex shape defined by points that can be edited in vector mode, or simply colours in the touched pixels in bitmap mode.

Fill Tool. The Paint Bucket, or Fill tool, allows colouring an area on the canvas. In vector mode, this must be a defined shape or space completely surrounded by defined shapes. In bitmap mode, this can be any area and will fill in any contiguous area by colour. So if you click on the empty canvas, the entire background will fill in. If you select an outline of an object, that whole outline, in so much as it's a single exact colour, will fill in, but any difference in colour from the selected origin for the fill will not be affected. The Fill tool applies the currently selected fill colour (or gradient) to the area it applies to.

Line Tool. This tool allows you to create lines of the colour and thickness selected by the Outline Colour and Outline Width properties. In the vector mode, this will create objects defined by points, and by making multiple lines, you can connect into a single object, which can alter their character. In bitmap mode, the Line tool simply fills in the pixels touched by the line to the outline colour. In vector mode, if a user holds the Shift button while creating a line, it will snap to 45-degree angles.

Rectangle Tool. A creator can make rectangles and squares using this tool, which will have the current Fill, Outline, and Outline Width properties applied to it. Simply click where to start and drag out to create the size and shape you wish. In vector mode, this will create a shape object that can be edited with each of the four default points having the Pointed property. In bitmap mode, it will simply fill in the pixels affected. In either mode, if the creator holds down the Shift key, they will create an exact square; if they hold down the Alt key, they will create a symmetrical shape centred on the initial click.

Circle Tool. Just like the Rectangle tool, this tool functions the exact same way but makes circles and ovals instead. In vector mode, the shape will have points with the Curve property instead of pointed.

Text Tool. This tool allows the user to create text elements in their art. Simply click to start creating text, then you'll be able to type in the text you want. The text will have the fill colour currently selected and will also use the Font property, which allows you to choose from a list of available font types. The text will have a box around it with points at the cardinal and corner points that you can click and drag to resize. When done typing, simple click on the canvas to complete the text to set it into the pixels in bitmap mode. In vector mode, the text remains a distinct object you can later select and edit, including changing the text.

Eraser Tool. The Eraser tool allows one to erase, clear, or remove shapes and colours from the art. Simply select the tool, and if desired, change its size with the Eraser Size property and then you can click on the canvas or "draw" with the Eraser tool and everything that it overlaps will be removed from the art. In vector mode, this will remove portions of shapes or other objects (except text) and will alter their vector instructions, adding points to alter them to exactly how the eraser affected them. In bitmap, it will simply set any pixel it overlaps back to transparency.

Reshape Tool. This is perhaps the most powerful and complex tool to learn and master. The Reshape tool is only available in the vector art mode. It allows the user to select a shape and see the various points that define it. With the Reshape tool, the creator can select individual points or, by clicking on the shape's lines, insert a new one. Selected points can be switched between curved or pointed types, be moved, or if curved, have their handles moved to change the nature of the curvature of the lines extending from it. It's impossible to describe in words quite how one can manipulate shapes using this tool. It must be experimented with and practised, but it is well worth the practice. If one holds Alt while manipulating a handle, it will act separately from the other handle; otherwise, the handles work in tandem to create symmetry. Be warned that splitting the handles with the Alt key is permanent, and the handles will never again adjust symmetrically.

Art Tool Properties

When you're using the art tools, a number of properties will apply depending on the tool selected. The shapes or brush will have a Fill property, which defines the colouring of the shapes they make. To choose the Fill property, click on the Fill button above the tools. This will display a colour panel, where you can choose a colour (also known as hue), saturation, and brightness (also known as lightness). The combination of these three properties will determine a specific colour. In the bottom left corner of the panel, one can choose Transparency, or in the bottom right, one can use the colour picker to grab a colour from the canvas to use. In addition, one can choose to use either a solid colour or a gradient by selecting one of the four options at the top of the colour panel. A gradient will smoothly and gradually blend between two chosen colours. If using a gradient, two colours will appear in the colour panel, and you can click on either one to adjust them or use the Swap button to exchange the colour positions. The colour panel is used in many places in Scratch but may have some slight changes depending on the context.

Any Shape or the Line tool can have Outline properties. Clicking the Outline button (next to the Fill button) will bring up another colour panel to choose a colour scheme for the outline applied to the shapes or lines being made. Again, you can also select Transparency, or no outline, or use the colour picker to grab a colour already in use on the canvas.

If an object has an outline, it can have its outline width set. This is also what is used to determine the width of any lines drawn with the Line tool. The number to the right of the Outline button is nominally the number of pixels thick an outline will be drawn at. If you resize an object, its lines remain the same thickness, so you may need to adjust this number to get the same proportions as you change sizes of shapes, but resizing in code will maintain the ratios and proportions of costumes.

If you select the Brush or Eraser tool, you can set the Size property for them. The number to the right of the outline width is the brush or eraser size nominally in pixel width. The shape of the brush or eraser is always a circle, with this number giving the diameter of the tool as it will apply on the canvas.

Compositing

In vector mode, you can alter the composition of your designs through the use of layering and grouping. Layering controls how shapes are drawn on top of or behind each other. After selecting any object or group of objects, you can use four buttons to control their layering. Above the canvas you'll see Front, Back, Forward, and Backward buttons. These will move an object either forward (toward the user, above other objects) or backward (away from the user, or behind other objects), either one layer at a time, or go right to the limit of

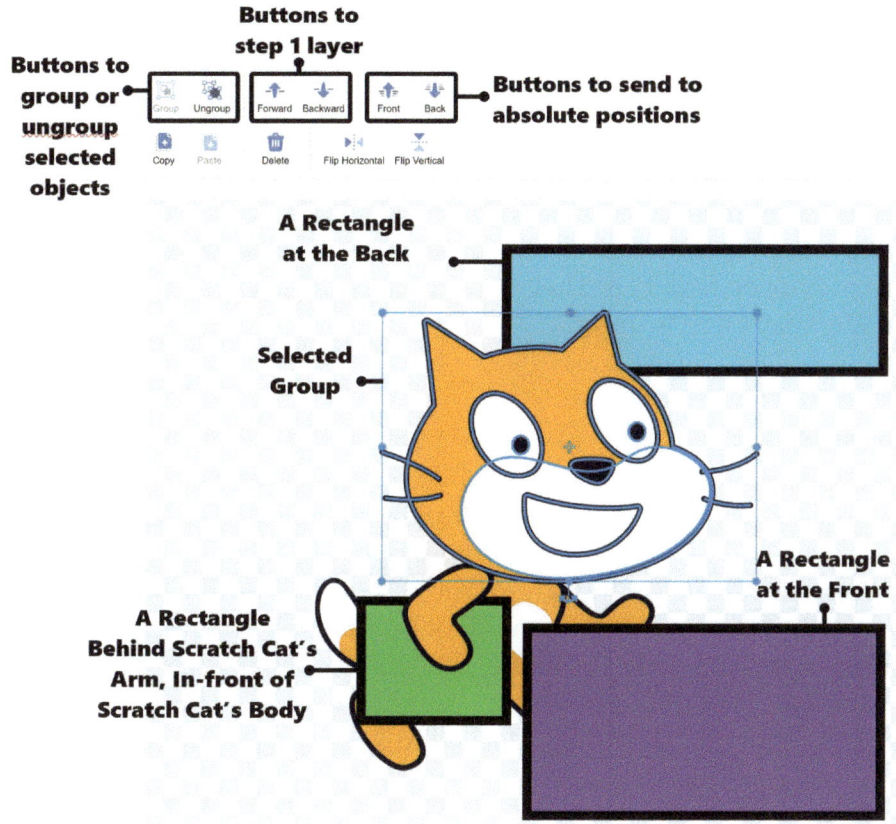

Figure 4.16 Exploring the composition tools. Objects on the canvas can be layered in front or behind each other using the Layer buttons. Objects can also be grouped together to maintain their relative sizes and positions.

the order. Layering is vital for images to display correctly, so for example, you can make sure eyes are visible in front of (or forward from) a head.

Sometimes shapes may be integrally connected with other shapes but still need to be moved in parallel. It can be difficult to select all the objects every time you need to reposition something. To make this easier, you can create groups of objects. Objects that are grouped can be selected and moved as a single collective and will stay in proportion to each other. Groups can still be edited by ungrouping them and editing the individual components again normally. By grouping we can try to simplify our lives by creating more easily animated sprites, for example, linking clusters of objects such as eyes and head together, then that group can more easily be moved or rotated as needed without reselecting all the components or adjusting them separately. Groups can also be grouped again into groups, so in some sprites you may have to ungroup, select again, and ungroup again to drill down to an individual component you want.

Undo

People inevitably make mistakes. If you want to go back on a change you've made in the art editor, there are Undo and Redo buttons above the canvas. You can generally undo through a sequence of changes. You can also use the standard CTRL-Z keyboard shortcut. The Costume tab Undo is separate from the general Edit menu Undo that can help you undo code or sprite management changes. However, all the Undo features in Scratch are fairly limited and should not be relied on. They may or may not be able to undo the change you made, so caution is urged.

The Sounds Tab

The Sounds tab is relatively simple in comparison to all the depth of the Costumes tab but still offers some interesting options. Through this tab you can add sounds to a sprite (or the stage). You can find the Add a Sound button in the bottom left corner. Clicking on it will take the user to the Sounds Library to select a sound to add to the sprite. If hovered over, it will extend and offer other options, including recording a sound, getting a random sound, or uploading a sound. The sounds attached to a sprite show up on the left-hand

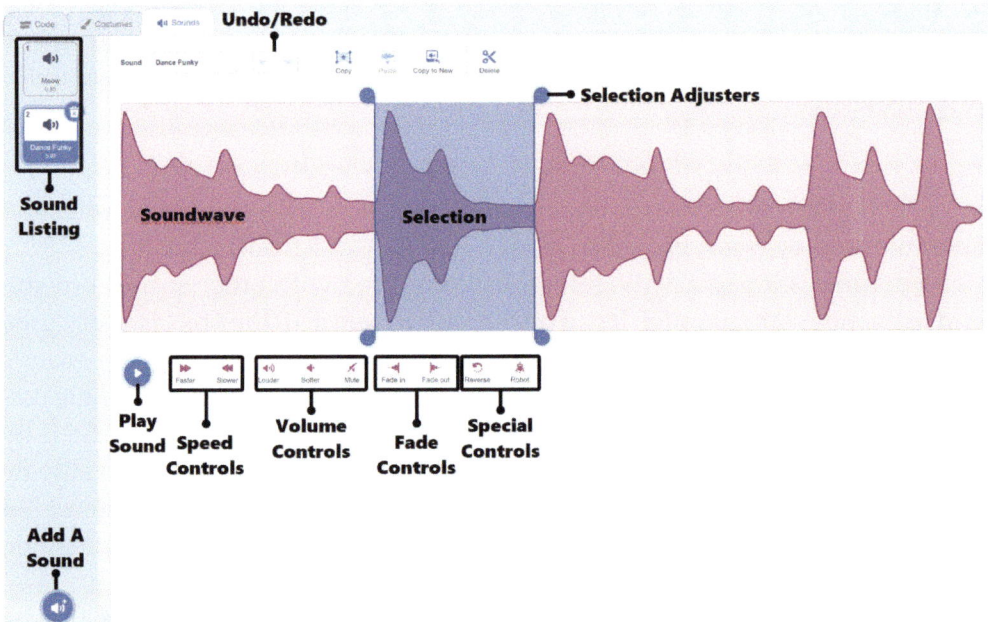

Figure 4.17 Scratch's built-in sound editor can help you modify and explore sound and music and gives a great visualization of sound waves!

side of the editor in a Sounds List; just like costumes, each one has a name and a number, and they can be clicked and dragged to rearrange, although there is no Next Sound code block, so the order doesn't really matter.

The Sound Library works similar to the Sprite Library. Scratch comes with a wide array of sounds that you can instantly add and use in your project. We can't see sounds to browse through them, but the names are fairly good descriptions of what the sound will be. Importantly, each sound tile has a Preview button you can play by hovering your mouse over it. If you click on a sound, it will be added to your sprite, so remember that you only need to hover your mouse to hear it. At the top of the Sprite Library, there are categories that you can click on to reduce the list just to sounds of that category.

In the workspace, we can see a minor sound editing program. Whatever the current selected sound is, we can see a waveform for the sound. Select time frames on the waveform by clicking and dragging to select an area of it. This selected area will be where any effect we select is applied to; alternatively, without an area selected, the changes will be across the whole sound. To unselect an area, just click anywhere on the waveform. To change your selection, the tabs at the top and bottom can be clicked and dragged to slide the start or end of your selection forward or backward.

Below the waveform view, we have a few editing tools. First, a Play button to hear the sound with whatever changes we've made to it. Then there are buttons to change the tempo and volume, to mute a selection, to gradient-scale the volume to fade in or out, to reverse the waveform, or to "robotify" the sound. I'd suggest playing around with the tools to see what they do. Changing the speed of a sound also changes its scale.

Above the waveform, you can change the name of a sound, undo/redo changes, copy, paste, copy to a new sound, or delete a selection, shortening a sound by removing the selection completely.

Sounds are very data-intensive, so it's recommended for quickest loading to minimize any unneeded sound files in a project. You can delete sounds by selecting them on the left-hand side and then clicking on the trash bin that appears in the upper right-hand corner of their tile.

Projects

Now that you know what you are working with, what do you make? Scratch saves creations as projects, an obvious-enough construct. But what makes a project? A *project* is really a collection of digital objects, each with their own behaviours and assets (art and sounds), that act or interact in some way. These objects come in two varieties: firstly, the stage, an always-present, singular object that creates the area in which all other objects exist, and sprites, which can be present in any number (or none) and are customized to any appearance (or none) and operate independently. Keeping with the stage analogy, sprites are both the props and actors that bring a play to life upon the stage.

Each project saves all the objects it contains within itself, with each object saving its own code, art, and sound within it. Each project is a completely stand-alone experience, but you can copy objects, art, code, or sound from one project to another if desired. This is made possible with user accounts and the Backpack, which we will give a hands-on demonstration of in Project 1: Dino Dance Party. All the projects a user with an account creates are saved and listed on their My Stuff page, but users with or without accounts can save their projects to their local computers as well.

Sprite Listing

On the right side of the editor, regardless of which tab you're in, below the Stage Window you'll see a list of all the sprites (objects) in the project. Each sprite added to the project gets a little thumbnail added to this tiled listing. This listing allows you to not only see the objects added to the game but also click on them to make your actions in the three tabs apply to that specific object. Through the tabs you can alter their appearance (through costumes), behaviour (through code), sound (through sounds), or through the Property panel or code change their properties. In a project, each distinct object has its own appearance, behaviour, sound, and properties that determine what they are and what they do.

When you want to change things in your project, you need to make sure you are editing the right object. You can see which sprite is selected by a blue outline on the Sprite Listing thumbnail for that sprite, and if you're in the Code tab, you'll also see a little watermark or ghost of the sprite that's being coded in the upper right-hand corner of the coding workspace.

If enough sprites are added to a project, a scroll bar will be added to the Sprite Listing. Sprites can be dragged to rearrange them for convenience, and this has no bearing on the performance of the project. When a sprite is selected, it can be removed from the project by clicking on the little trash bin icon that appears in the upper right corner of the thumbnail.

Stage

Not every editable object is listed in the Sprite Listing. One remains separate and distinct – the stage. The stage is the background of the game, but it isn't just an image; it can have its own code and assets too! Learning to work with

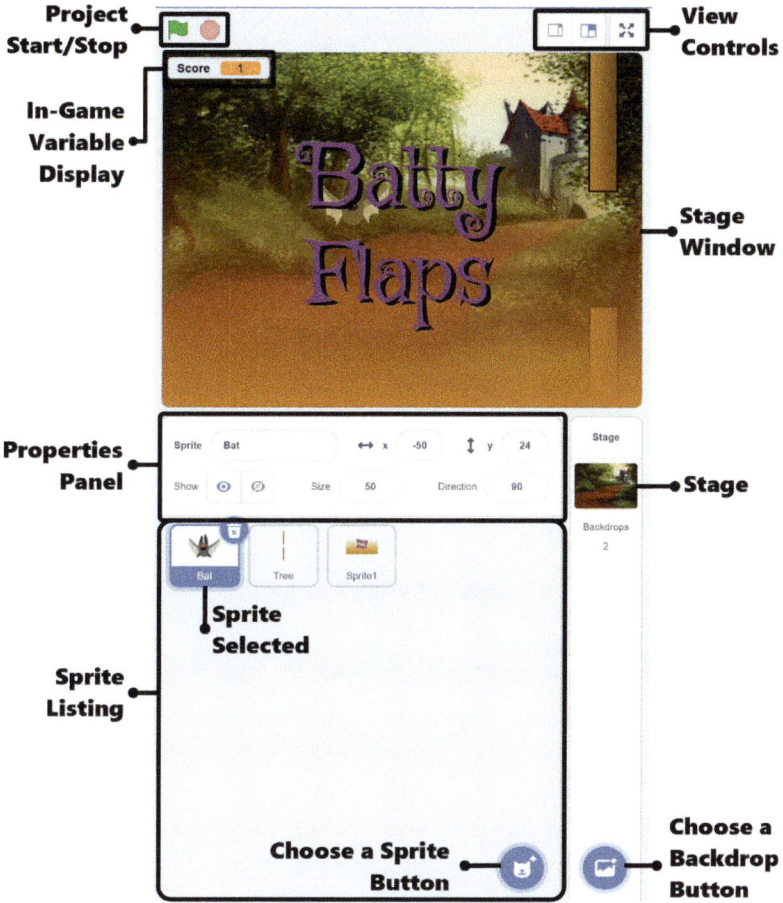

Figure 4.18 The right side of the editor, showing the Stage Window, Properties panel, Sprite Listing, and Stage.

the stage can be a handy skill to create more universal changes and controls in a project aside from individual sprites. The stage may be more abstracted than sprites and harder to conceptualize, but it's also simpler to work with. It has fewer properties than sprites since it cannot move, always sitting immobile in the background. As such, it has fewer code blocks it can use, but it can still be a very useful part of a project, despite this simplicity.

You can select the Stage on the bottom right-hand corner of the editor. The stage has its own appearance, determined by what backdrop it displays underneath all the sprites, behaviour, and sound, but it doesn't have editable properties other than the ones changeable through code. The stage is primarily there to let you "set the scene" in your project using backdrop images, but it can also be used for code functions as well, although you'll notice some code blocks are not available for the stage. Scratch has a whole bunch of backdrops built into the Backdrop Library that can be added to your projects instantly by clicking on the bottom right corner's Add a Backdrop button.

Stage Code

When you select the Stage on the bottom right-hand corner of the editor, you'll see a lot less code blocks available in the Code Library (on the left side of the screen, in the Code tab). It cannot move, and so no •Motion blocks are available. Only a few •Looks code blocks are available, and a few others are removed. While the stage may not be able to move, change many aspects of its appearance, have clones, or collide with sprites, it is a handy addition to the sprites. It can be a great place to handle more universal aspects of a project, such as background music, creating slideshows, determining what level

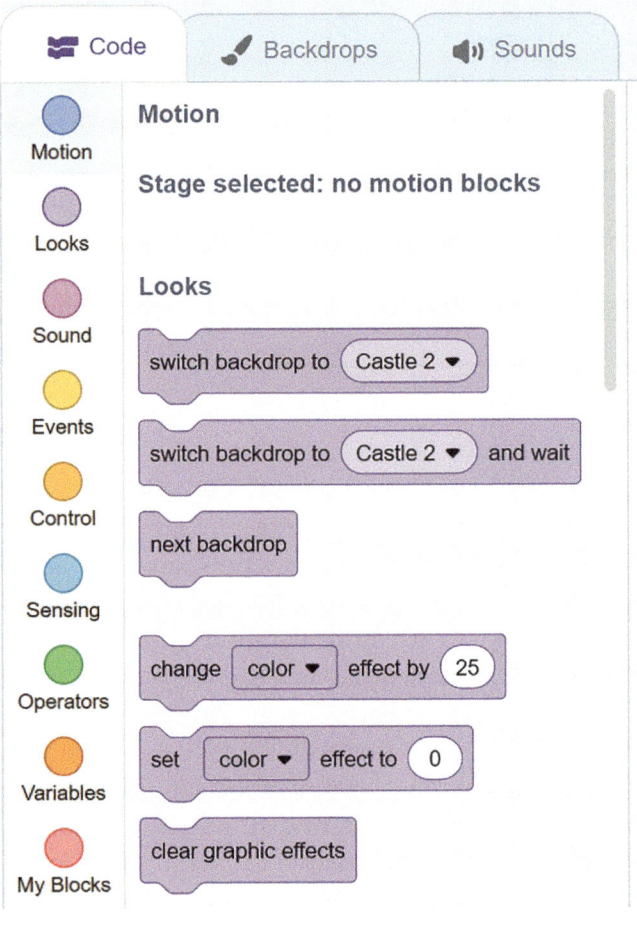

Figure 4.19 A look at the Code Block Library when the stage is selected. Notice the lack of motion blocks and costume-related blocks.

is being played, tracking data or variables, or creating menus and navigation. The biggest drawback of the stage isn't its code or coding options but rather the in-classroom experience. You'll find new students may end up selecting the stage instead of a sprite and not realizing their mistake until they can't find the code block you're instructing them to use. If a student can't find a code block, they've probably selected the stage and may need to move some code they've accidentally added to the stage to the correct sprite in order to fix things.

Backdrops

The primary purpose of the stage is to provide a background image. It does this through backdrops, which function like costumes do for sprites. The stage can have many backdrops added to it through the Backdrop Library, by creating backdrop art yourself in the Backdrop tab, or by uploading some. While the stage can have many backdrops, only one is shown at a time: the current backdrop. By switching the current backdrop, we can change scenes or slides in our project.

Stage Sounds

Just the same as sprites, the stage can have sounds associated with it and then called by code. In the Sounds tab, one can add sounds, either music or sound effects, to the stage by selecting them from the Sounds Library, recording, or uploading them.

Sprites or Objects

Now that we understand the overarching structure and components of Scratch, let's take a closer look at how sprites, the non-stage objects in projects, work in Scratch. You'll spend most of your efforts designing how these individual components in a project look, act, and sound. Each sprite gets its own graphics, code, and optionally, sound effects. We'll use code and the editor to change their properties to give them different states, appearances, and behaviours to bring projects to life.

Adding Sprites

All projects start with Scratch Cat as *Sprite1* so that first-time users have something to look at, and if they grab a code block, they'll have something that will react and move. To add other sprites to the project, we can use the Add a Sprite button in the bottom right corner, the one that looks like a cat face. If you click on the button, it takes you to the Sprite Library, a collection of art built into Scratch that can instantly be used and worked with. If, instead of clicking on it, you hover your mouse over it, you'll get other options. Besides the Sprite Library, you can create your own art with the paintbrush icon or upload your own art with the upload icon.

The Sprite Library contains over a hundred premade objects you can add to a project with the click of a button. Scrolling down through the list will show every available premade object. If you hover your mouse over a sprite in the Sprite Library, it may animate. This animation will show you the different costumes that come automatically built into that Sprite. Some may have walk cycles, dance poses, expressions, or different styles, while some have no additional costumes and will not animate.

Using the Upload function, if you upload a. jpg,. png,. bmp,. webp, or. gif file, you'll get pixel art, while uploading an. svg file will get you vector art. We talked about the difference between pixel and vector art in the Costumes/ Backdrops tab section in this chapter earlier. Uploading can allow you to access all sorts of art, but you may want to talk about fair use and plagiarism with your students regarding using the Internet to get art. You may also find that uploaded art is very rarely convenient to use in Scratch, as separating parts of images from their background is a tricky process. Learning to use the built-in art tools that Scratch provides in the Costumes tab can be an incredible learning, growth, and creativity opportunity for students (and teachers), so I highly recommend getting familiar with it. We designed a number of

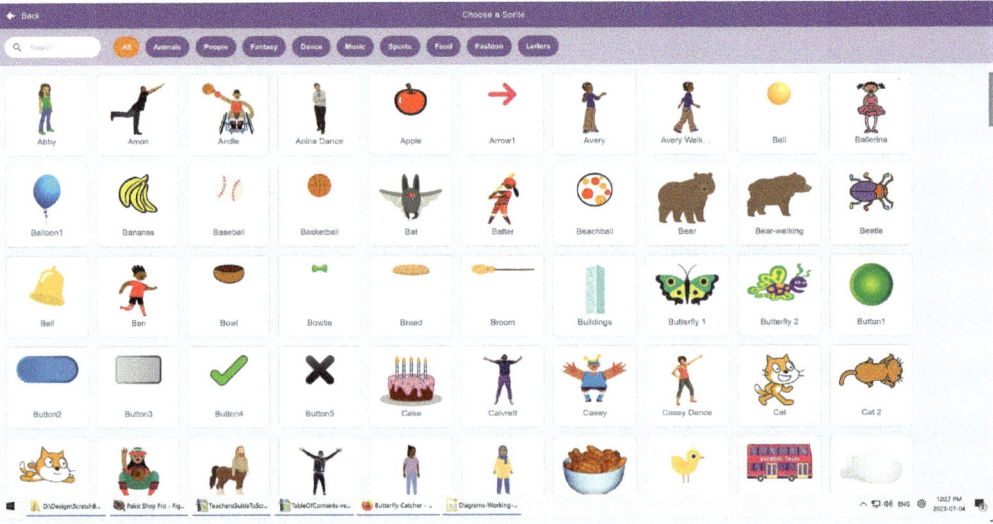

Figure 4.20 A glimpse of the Sprite Library – premade art available in Scratch. A wide array of art is available to work with, so you can jump right in without having to make your own art, some including animations; just hover your mouse over a sprite to see!

projects in this book series to help you explore working with the art tools in the Costumes/Backdrops tab to help with this.

Arranging and Deleting Sprites

The thumbnails of sprites in the Sprite Listing can be rearranged. They are simply listed in the order they are added to a project but can manually be rearranged by clicking and dragging them to your convenience. If you rename a sprite or change its costume, those changes will be reflected in the Sprite Listing when confirmed. If you no longer want a sprite in your project, select it and a trash bin icon will appear on the upper right-hand corner of the thumbnail. Clicking the trash bin will delete the sprite along with all its code, costumes, and sounds and its instance in the Stage Window. If you accidentally delete a sprite, you can undo that deletion by going up to the blue bar at the top of the screen and finding the Edit button, clicking it, and selecting Restore Sprite. You can also use CTRL + Z to undo the operation, just like most programs. Just keep in mind that if you've done other work, the option to undo may no longer work for the problem you want to undo, especially if you've switched tabs.

Properties

Every sprite has a number of inherent qualities assigned to it, called properties. These properties help define the sprite. Directly below the Stage Window and above the Sprite Listing in the editor, you can see a Properties panel, where a number of these are listed. A sprite has a name, X position (where it is left/right in the project), Y position (where it is positioned up/down in the project), visibility state (visible or invisible), size (a percentage of how large or small the sprite is drawn in the project compared to the size it's drawn in its costume art), and direction (what direction it faces for movement 0 = up 90 = right, -90 left).

There are other properties not listed in the panel, such as its costume (the graphic currently being used to represent the sprite if visible), layer (where it is drawn in the game, determining what is in front or behind it in the drawn order), ghost (level of transparency the sprite is being displayed with), colour (the hue shift applied to the graphic being used to display the sprite), and others that aren't listed but can be influenced by the code.

By changing properties, we can change how the sprite exists in our project's world, making it move around, or change its visibility. Some properties may affect how code is interpreted. The size of the object will affect what it touches. Direction determines its rotation (depending on its rotation-style property), or at least where a Move command will move it toward. Visibility will also determine whether it touches other objects at all or if its Say or Think blocks will display.

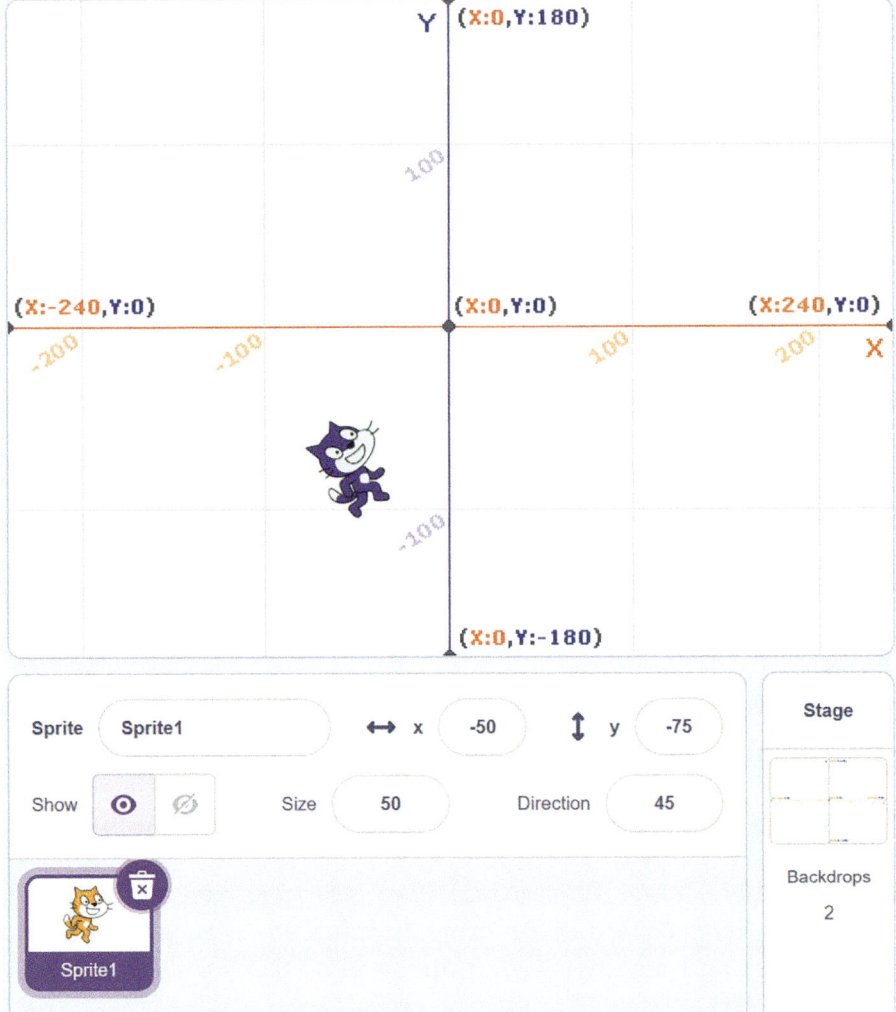

Figure 4.21 The Property panel can help you set an object's current properties or tell you what they are, which can be a very helpful aid when troubleshooting! No property controls are given for graphical effects, but this particular Scratch Cat has had its colour shifted. The Properties panel is very useful but is not exhaustive.

Code

Coding is what Scratch is all about! Sprites are the workhorses of projects in Scratch, and code is how we get them to do stuff. Each sprite in Scratch can have its own code, so that each one can act in its own way. The coding workspace displays the code for the currently selected sprite. Code can be assembled by dragging out and assembling code blocks from the Code Block Library on the left-hand side of the screen when on the Code tab in the editor.

Costumes

Each sprite is displayed on the screen in a certain way at any given time. This is achieved through a number of graphical properties but, most importantly, the current costume of the sprite. Each sprite can have multiple costumes but only wears one at any given time. Sprites can change their costume to change how they look. This can make them radically different, or it can just be a small change or even a frame of animation. Often, sprites will have multiple costumes for different poses, and many can switch between these to give an animation effect, such as walking, waving, spinning, or showing an emotion. Scratch has not only tools for coding but also art tools, so you can make or edit your own costumes for sprites, as well as use costumes via the Sprite Library, Costume Library, or you can upload them.

Sounds

Just like costumes, sprites can have many sounds associated with them. Songs or just short sound effects, they both work the same. Just like costumes, sounds must be added to a sprite for it to be able to use them. Sounds are added in the Sounds tab with the Add a Sound button and can be selected from the Sound Library, recorded from the microphone (if one is connected to the computer), or uploaded. Once a sound is added, you can set it to play the sound when needed through code using the ●Sound category code blocks.

Next Step

Now that you've been introduced to all the various components of Scratch and how to work with them, let's get to building our own projects! The following six chapters will each instruct you on how to build projects that will introduce different features and concepts, and review your learning as you progress through them. They are arranged in order of complexity and difficulty, so you should start with the first and work your way to the last. Each chapter will help scaffold learning toward the more advanced concepts.

5

Defining Beginner Scratch

It's generally easy to define a beginning – it's the part with nothing before it – but here we aren't just looking at the start of things but the mindset and approach we want to take with learners early on the path to coding. What does beginner Scratch look like? What does it include? What does it not include? What is the focus, and what are the goals we should have in mind for beginners? Every learner is unique, so of course there aren't strict definitions we can rely on, but we can have a mindset for how we approach the topic. Later in the book, you'll find I've provided additional challenges for the more advanced students you might have, so here we'll stick to generalities.

Beginner Scratch Attitudes

Beginner Scratch, of course, starts right from knowing absolutely nothing about coding. It's the period of the most discomfort, confusion, uncertainty, and doubt, but it's also the period of great wonder, excitement, discovery, and growth. The attitude I expect students to have is, "I hope we can build it". We want to encourage discovery and imagination while tempering our expectations with the doubt and natural ignorance we start from.

The focus here is on core learning – where things are, what things do, what we need to do, how it works, on the broadest and simplest terms. Projects should be both small and simple. We stick to basic actions, use a relatively

DOI: 10.4324/9781003399018-5

narrow range of techniques. We use repetition to ensure ideas are solidifying in young minds and that techniques will be remembered.

What we don't deal with are complex interactions, complex formulae, recursive functions, custom blocks, dynamic or adaptive programming, or projects that are too large or too complex with object types. Keeping things simpler helps keep what we do be clearer, easier to follow and comprehend, and easier to remember.

Beginner Goals

When beginners are ready, they'll start adapting programs, adding more, tweaking things. When they start trying to reach further, you'll know they've reached the comfort level for intermediate Scratch. Ideally, our beginners will reach our goals of:

Familiarity. Know the layout and components of Scratch and how to find things, how to work with blocks, save and load projects. They'll properly identify where they are and where to go, will have used all three tabs, and know the difference between the stage and sprites.

Confidence. Know they can make things in Scratch and not fear trying (too much).

Moving. Are confident moving sprites with both centric and absolute movement. They'll know the difference between ●[Go] and ●[Glide].

Animating. Know sprites have costumes and the stage has backdrops; will know how to load them and to change them with code.

Coding Fundamentals. Understanding the importance of sequence, needing to trigger code with an event, the ability to have concurrent events, using repeating events, understanding the project-object-property model.

Events and Loops. Can use the ●[When Sprite Clicked], ●[When ($) Key Pressed], ●[Forever], and ●[Repeat (#)] blocks to get things to happen when and for how long they want.

6

Beginner Project 1: Dino Dance Party

What This Project Is

In our first project, we take things nice and slow. We'll be making a simple music-and-animation show using premade art. The project will consist of a few dinosaur characters, and the player will be able to press keys to have each one animate and play music. I'll walk you through how and where to find all the different tools in Scratch needed to make your first project. It should take less than one hour for you to complete and will provide you a good sense of how to make simple projects and find the major features of Scratch.

What We're Learning with It

We'll be introducing all the basic necessities of working with Scratch. Navigating the different features of the interface, taking a look through the various built-in libraries of assets, adding objects (sprites) to a game, using player inputs and basic coding. Specifically, we'll be working with:

Stage Code. Adding code to the *stage*, a somewhat-hidden feature to new users, and one that can cause some hiccups when teaching. We'll explain how and why we use this, and the important way to avoid trouble in classroom instruction.

DOI: 10.4324/9781003399018-6

Figure 6.1 The Dino Dance Party project finished.

Sequences. We'll make some basic sequential code to have our objects (sprites) react and follow instructions in proper order.

Events/Triggers. Our program will respond to multiple events, including user input; we'll see how different events trigger, and use those to create an interactive experience.

Animation. Using built-in sprites, we'll use their different costumes to animate them in two different ways.

Timing. We'll explore the ●**[Wait (#) Seconds]** and ●**[Repeat (#)]** code blocks to match our sequences to music to time reactions and events.

Sound. We'll use built-in music samples to create looping background music, as well as getting sound to play on command.

Building It

Step 0: Create Your New Project
Make sure you're logged in to Scratch, then click **Create** to begin a new project!

Step 1: Adding a Background
Our first task is to set the stage, literally. The stage is where our game or project takes place. It's the background image behind everything, and it can

also include coding. We're going to set the mood for our project by selecting a background image and some music using code.

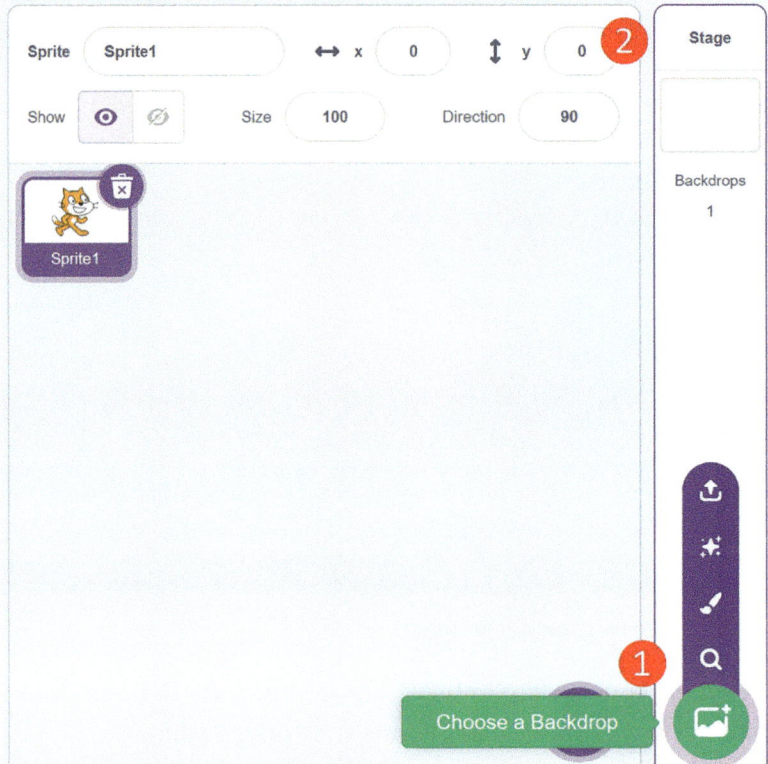

Hover your mouse over the **Choose a Backdrop** button in the ① lower right-hand corner of the screen. It will turn green and display four options on top of it. By clicking on the main button (or the magnifying glass directly above it), we reach the library of built-in stage backgrounds. For this project, we'll use the **Concert** backdrop. By clicking on the backdrop we want, it becomes the current backdrop for the game, and it moves us back to our main editor view.

Object Selection

Now, an important thing to note is that we now have the Stage selected. You can tell that the stage is selected by the blue outline around the thin tall Stage portion of the screen in the bottom right. When we start a new project, Scratch Cat is selected, and the blue outline is around the little thumbnail of Scratch Cat in the Sprite Listing that belongs to the Stage Window. Because we clicked the Choose a Backdrop button, we've switched to the stage. The stage can have its own costumes (backdrops) as well as its own code, but it has more limited options for coding. It is common for students to not realize what object or stage they have selected when they are getting to know Scratch. Make sure you get into the practice of telling students what sprite, or stage, to select and add code to, or they may be working on the wrong thing. You may only notice this when you tell the students to get a block that a student says they can't find, an indication they're on the Stage and can't see certain blocks. Having the right object selected is key, so make sure you call it out. Another handy feature to notice is the little ghost image of the currently selected object in the upper right corner of the coding portion of the screen.

Now that we've set the backdrop, we can add in our first code. Let's make a song play in the background to enhance our concert mood. Click on the **2** word "*Stage*" on the right side of the editor so we can code some behaviour for our background.

First, we need a yellow •**Event** block so Scratch will know when to trigger this code. Here we'll use the first code block under •**Events**: •**[When ▷ Clicked]**. This ensures the code will run as soon as the project starts. Under this we'll add an orange •**Control** code block: •**[Forever]**. This code block makes any code put inside it run repeatedly forever. By putting our music code blocks within a •**[Forever]** code block, we can ensure the music will repeat (or loop) as long as the project is running – perfect for background music.

Now, to play some music, we first have to add the music we want to the project. For this, we'll need to go to the **Sounds** tab at the top left corner of the screen above the code block list. These tabs switch between an object's code, art, and sound assets. By selecting sound, we'll see what sound effects are loaded and associated with our selected object. They'll be able to be used by this object, but not by others.

To add a sound, once you're in the Sounds tab, click on the **Choose a Sound** button in the bottom left of the screen. Similar to choosing a background, this lets us browse the built-in Sound Library in Scratch. Because we're looking for a piece of music rather than a sound effect, we can click on

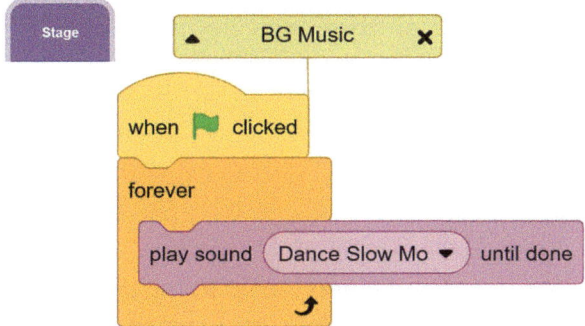

the *Loops* filter button at the top to narrow down our options. We're going to select the **Dance Slow Mo** loop. If you want to preview it, you can hover your mouse over the Play button for that sound. This sound clip will play in a seamless loop for our project, giving us some suitable music for dancing dinosaurs. After adding it, we can return to the **Code** tab to call it in our code. Last step is adding the ●[**Play Sound (***sound***) Until Done**] code block to our ●[**Forever**] loop and selecting the **Dance Slow Mo** clip as the sound it will play.

Now if you click the ▷ button to test-run the project, you should hear the music play and loop seamlessly forever in the background. I suggest clicking the red Stop button to stop the project so you don't get too sick of the music while we build the rest of the project.

Step 2: Adding the Dinosaurs

Now that we've taken care of the setting, we can now move on to the stars of our show – the dinosaurs.

To add a new sprite to our project, you need to click the ❶ little cat-faced **Choose a Sprite** button in the bottom right of your screen, just to the left of the **Choose a Backdrop** button. Click on this and it takes you to the built-in library of sprites. Scroll down to see a number of options for dinosaurs. We'll start with *Dinosaur1*, and then we'll repeat the process to add *Dinosaur2, 3,* and *4*. We've now got a script of dinosaurs on our stage. In our Stage Window, we can click and drag each dinosaur to position them where we would like.

Before we get to coding our dinosaurs, we're going to remove Scratch Cat from this project. We're not going to be using it in our project, so we want to delete it. If you select Scratch Cat, you'll see its thumbnail, labelled "*Sprite1*", below your Stage Window get highlighted in blue. On the ❷ top right-hand corner of the thumbnail, you'll see a trash bin icon. Click on that icon to delete that sprite. Now we've just got our dinosaurs onstage – let's start coding them!

Step 3: Animating the Dinosaurs

With our dinosaurs added in, we can now start coding them to act. Our first job will be to give them an animation cycle that will always run so they'll always be rocking out to the beats of our background music. To start with, we'll select *Dinosaur1* and get our code working for it first.

With *Dinosaur1* selected, we'll go to our ●**Events** code blocks and grab a ●[**When** ▷ **Clicked**] code block. We want our dinosaurs to start moving right away when the music starts. Next, we'll go to the orange ●**Control** code blocks and add a ●[**Forever**] code block so that they'll keep dancing as long as the project is running. Next, we'll need to go to the purple ●**Looks** code blocks and get a ●[**Next Costume**] code block and put that in our ●[**Forever**] code

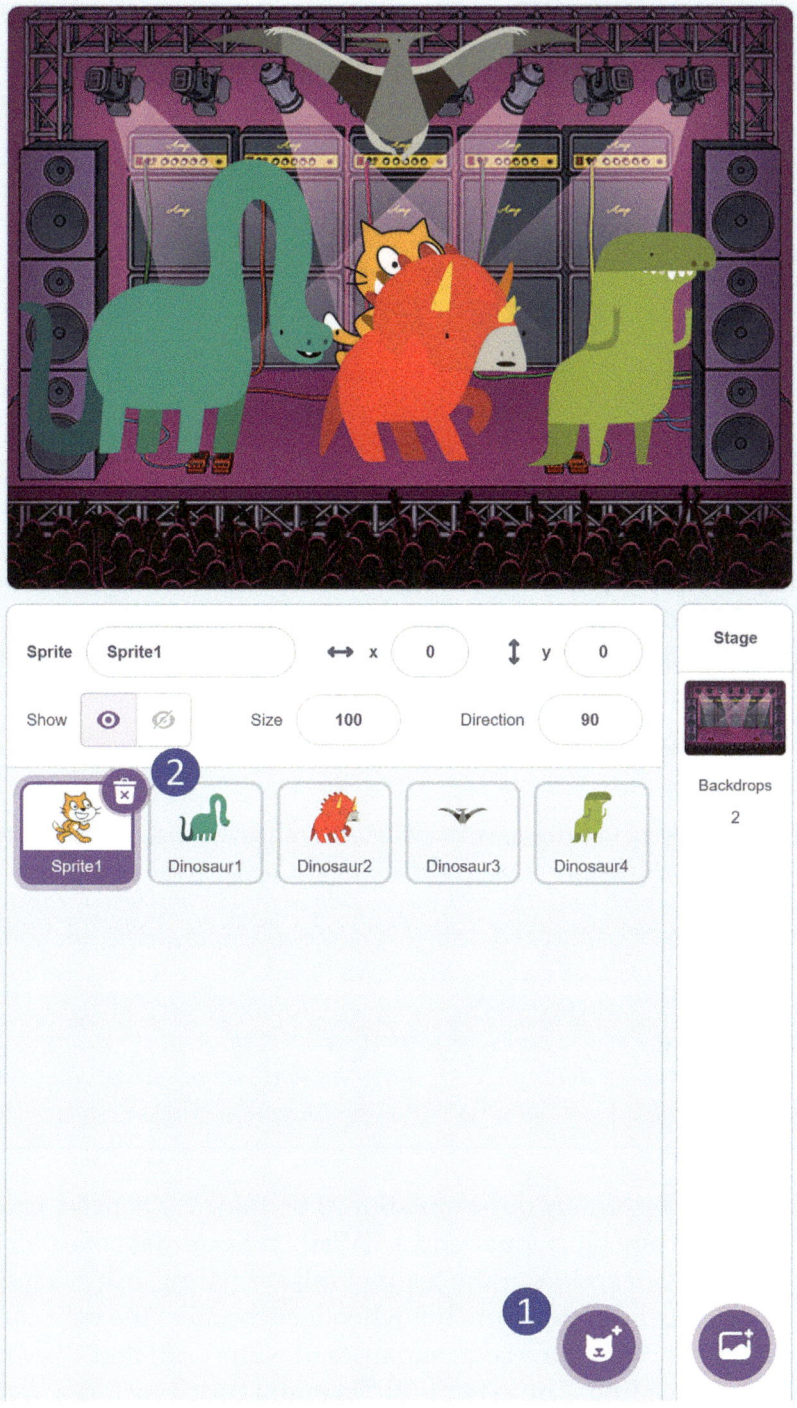

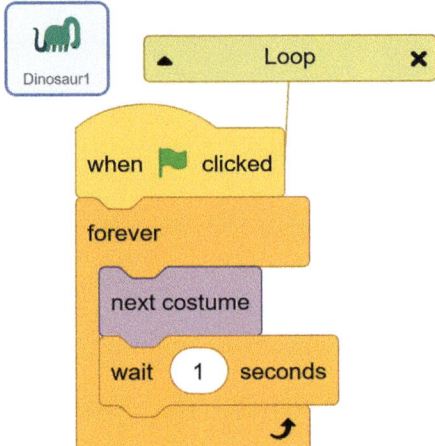

block. The ●[**Next Costume**] code block will switch what graphic is being drawn to represent our sprite. To see all the costumes that a sprite has, you can switch to the **Costumes** tab, and their costumes will be listed down the left-hand side of the screen. Some objects only have one built-in costume; others have many. Thankfully, all our dinosaurs have a few costumes that will work great for dancing.

Animation Speed

Now with our ●[Next Costume] block added, we can give a test run of our project. Click the ▷ and see how Dinosaur1 is animating. It's a little fast, isn't it? This is because of how Scratch runs. The code we've written for our dinosaur is constantly running. Every 1/30th of a second, Scratch runs the code and updates the game. This is really handy because it allows us to have things change rapidly, giving us a responsive medium. By running our code every 1/30th of a second, Scratch is animating our game at 30 frames per second, which is what most TV runs at. The problem here is that our dinosaur, then, is getting told to change costumes every 1/30th of a second!

So let's try slowing down our animation. For that, we'll head back to the orange ●**Control** code blocks and add a ●[**Wait (#) Seconds**] code block. When we drop that into our code, our dinosaur is still animating, but at a much more suitable rate. Here the code is run, but when it encounters the wait code block, it stops reading the rest of the code and instead waits until that 1-second timer is up before proceeding. The ●[**Wait (#) Seconds**] code block is a very handy tool for timing things out. You can change the amount of time the ●[**Wait (#) Seconds**] code block waits, including portions of a second. Feel free to experiment, but I think 1 second works well for the tempo of this music. Again, you can click the ▷ to play-test the project, and the **Stop** sign to stop it.

Step 4: Copying and Pasting Code

We've got one of our dinosaurs dancing now, but what about the rest? We could repeat the process for each of them, but thankfully, we've got some handy tools to make life easier. In Scratch, there are multiple ways to copy and paste code you need to reuse, so let's try them out!

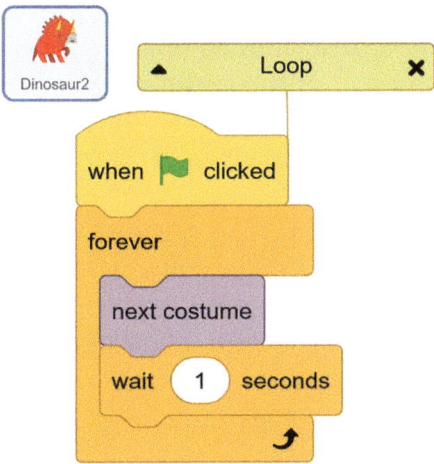

1. **Ctrl + C and Ctrl + V**

Our first method is the most familiar. Try selecting the ●[When ▷ **Clicked]** script of code blocks by clicking on it with the mouse and moving it a little. It will move the whole code with it. Now, press Ctrl + C. This should copy it to memory on your computer. By selecting *Dinosaur2*, it should give you a blank slate with no code. Try pressing Ctrl + V and you should see a copy of the code blocks appear for this dinosaur.

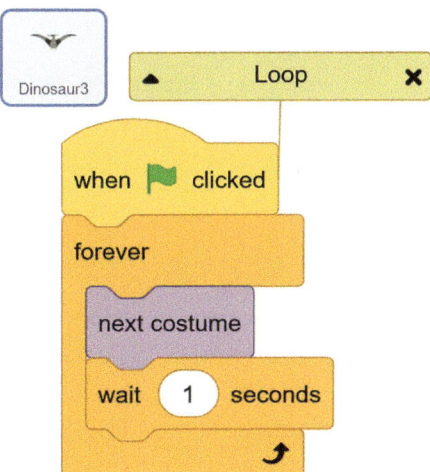

While this method is very familiar to adults and easy to remember, it can be tricky for younger students and is also not very obvious about what you might be copying or pasting.

2. Drag and Drop

Our second method is very visual and tactile. Here we can select the code block script we want by clicking on the ●[When ▷ Clicked] block and moving it. To copy and paste code to another object, simply drag the script of code blocks over to the thumbnail of the sprite you want to paste them to. Now the trick here is that it will detect position based on the mouse, not the whole size and shape of the code block script, which will probably touch multiple sprites. If you click on the code block script close to the top left edge of the ●[When ▷ Clicked], you'll be able to see where the mouse is easier. When you have it over a sprite, that sprite thumbnail will wiggle slightly to tell you that it is the current target. So when you move the blocks and see *Dinosaur3* wiggle, let go of the mouse. The code blocks you selected will go back to where you got them, but if we select *Dinosaur3*, we should see a new copy of the code blocks there. If not, try again.

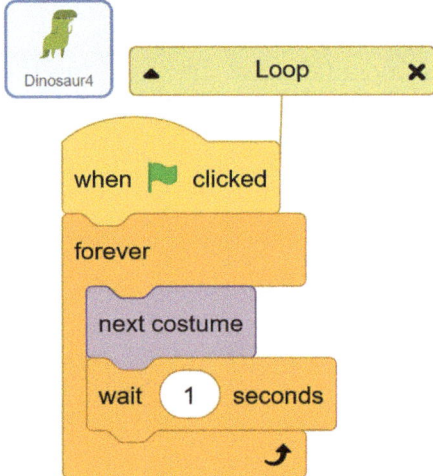

3. The Backpack

Lastly, Scratch has a built-in feature to help reuse code that's easy to miss. At the bottom of the editor (only when you're signed in), you'll see a **Backpack** button. If you click on this, it'll expand a little area that you can drag code block scripts into it and make a copy of them. You can also click and drag out any code block scripts already in there. This will allow you to copy and paste code blocks from one sprite to another, but the Backpack is saved to your account, so it offers you an option to copy and paste between projects even, making it a very handy tool to know about. Try it out by dragging the ●[When ▷ Clicked] script to the Backpack, and then select *Dinosaur4* and drag a copy out.

You've now explored all three copy-and-paste methods in Scratch and all four Dinosaurs have their dance animation cycle.

Step 5: Brontosaurus Beatboxing

To make our Dinosaur Dance Party interactive, we're going to allow the player to play additional musical tracks on command. To start, let's select *Dinosaur1* to add this feature to.

First off, as always, we need to start with an ●**Event**. We'll go to the yellow ●**Event** blocks, and this time we'll grab the ●**[When [Space] Key Pressed]** code block. This event triggers whenever the player presses the space key on their keyboard. Now we can see a white down arrow next to the word "space". This is used to indicate there are other options. If we click on the word "space", we see a list pop up showing other keyboard buttons we can select. I recommend switching to the "1" key for *Dinosaur1*.

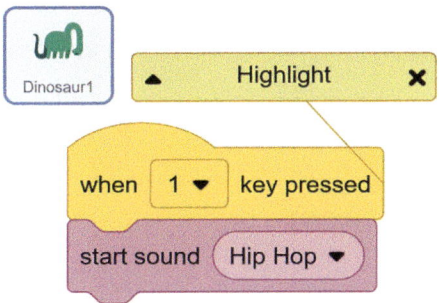

Now we're listening for player input or commands, but if you press the 1 key now, nothing happens because we haven't added any code to it. What we want is some more sound. So let's head to the **Sounds** tab so our *Dinosaur1* can play some music. We'll of course need to click the **Choose a Sound** button in the bottom left-hand corner, and we can narrow our choices by selecting **Loops** again. Here we're going to select the **Hip Hop** sound. Once added, return to the **Code** tab and add the ●**[Start Sound (*sound*)]** code block under our ●**[When [1] Key Pressed]**. Be sure to select the **Hip Hop** sound. Now you can press 1 and you should hear the Hip Hop sound play. This is a good start, but we're going to add more stuff before we bother copying and pasting it to the other dinosaurs.

In Book 3: Advanced, we've got a lot more music integration tips and techniques in our Point-and-Click and Scrolling Shooter projects.

Step 6: Highlight Animation

Currently, when our player presses the 1 key, they can get *Dinosaur1* to add in some beatboxing to the music. But they can't really tell it's *Dinosaur1* that's doing it. Let's add in some additional animation for *Dinosaur1* to highlight their added role when the 1 key is pressed and their music is playing.

Just like we did in the first animation, we're going to use costume changes to show an added level of energy and motion for the dinosaur we just called up to perform. We aren't going to be able to use exactly the same method, though, because their performance doesn't last forever. If we head over to the orange ●**Control** code blocks, we're going to see an alternative to ●**[Forever]** that will work perfect in this situation. Let's grab a ① ●**[Repeat (#)]** code block. Just like the ●**[Forever]** code block, this code block allows you to put code inside it that will be repeated (or looped); the difference is that this code block runs it a set number of times instead of forever. Inside our ●**[Repeat (#)]**, lets add a ●**[Wait (1) Seconds]** and ●**[Next Costume]** code block.

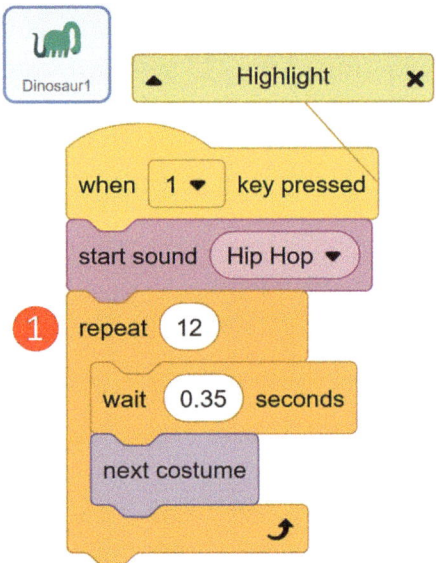

Try out your new animation by pressing the 1 key. You'll see that the animation works, but it doesn't seem quite right, does it? It doesn't time well with the music. Thankfully, we can play with the number of ●**[Repeat (#)]** as well as the number of seconds waited to time things out better. I find ●**[Repeat (12)]** and ●**[Wait (0.35) Seconds]** works out nicely to the music.

Step 7: The Music Is Bopping

So we've got a nice animation happening for our *Dinosaur1*, but what if we wanted to turn up the bass? Well, not literally; we won't get up to any sound editing for now, but what if we could tweak our animation to bring it even more to life and capture the energy of the music?

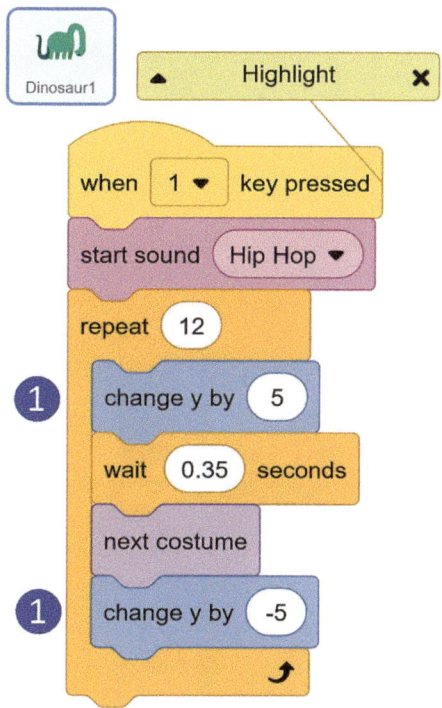

We haven't yet worked with any •**Motion** code blocks, so let's head up to the blue •**Motion** code blocks and explore another factor for our animation. Every object in Scratch has a position in the project, using an X-coordinate to determine its left/right position, and a Y-coordinate to determine its up/down position on a four-quadrant grid. Wherever you've put your dinosaurs, they'll each have their own X- and Y-coordinates to determine their position – meaning, where the computer should draw them. We can change their X/Y coordinates to move them. If we find the ❶ •**[Change Y By (#)]** code block, we can add one at the start of our •**[Repeat (#)]**, and another at the bottom of our •**[Repeat (#)]**. The top •**[Change Y By (#)]** code block, we'll make 5, and the bottom one we'll make -5. This way, each repeat the dinosaur will move up and then move back down to the original position. With the • **[Wait (#) Seconds]** code block between them, we make sure it doesn't happen too quickly.

Try out the animation by pressing 1. Our dinosaur now bounces to the music!

Step 8: The Whole Crew Is Jumping

Using what we learned in Step 4, copy the •[**When (1 Key) Pressed**] code block script to the other three Dinosaurs. This will give us the basics of what we need, but we'll need to go in and tweak each dinosaur so they have their own song and moves.

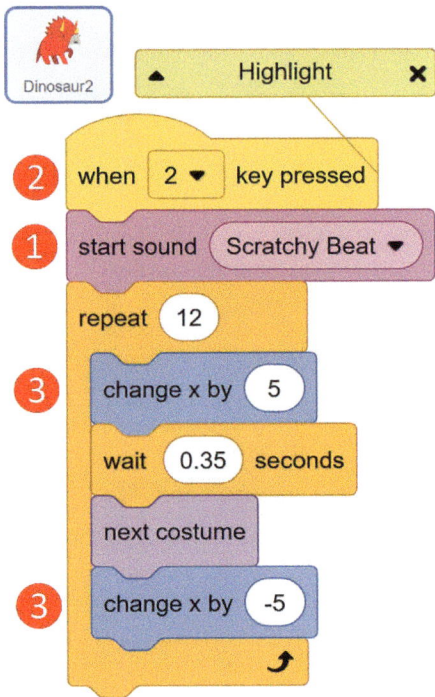

Dinosaur2 will need new music. Go to the **Sounds** tab and add in the **Scratchy Beat** sound. Now we can head to the code and switch the ❶ •[**Start Sound** (*sound*)] to **Scratchy Beat**. You'll need to switch the ❷ •[**When (1 Key) Pressed**] to the "2" key. You can also go to the •**Motion** code blocks and swap the ❸ •[**Change Y By (#)**] code blocks to •[**Change X By (#)**] code blocks so this dinosaur will move left/right instead of up/down, which I think matches the record *scratch* sound effects better.

For *Dinosaur3*, we'll change its ❶ music to **Dance Funky** and its ❷ key press •Event to "3". Now we'll also take the time to tweak the animation. We'll keep •[**Change Y By (#)**] since a flying dinosaur moving up and down really adds to the wing beat animation, but we'll tweak the timing a little. Let's try ❸ •[**Repeat (15)**] and ❹ •[**Wait (0.25) Seconds**].

Dinosaur4, we'll switch its ❶ music to **Drum** and ❷ key press to "4". We'll switch to ❸ •[**Repeat (14)**] and ❹ •[**Wait (0.25) Seconds**].

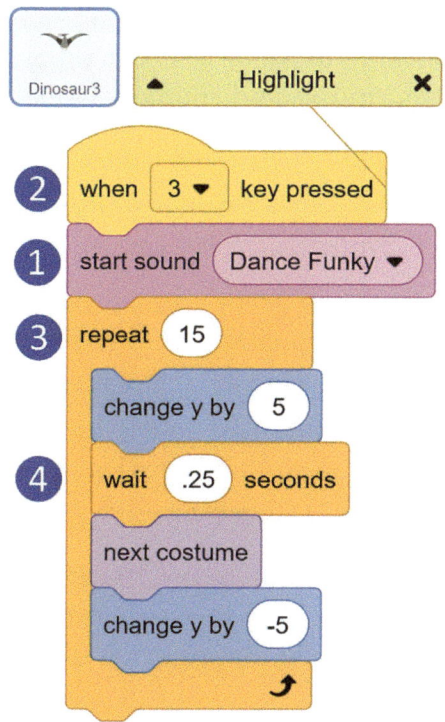

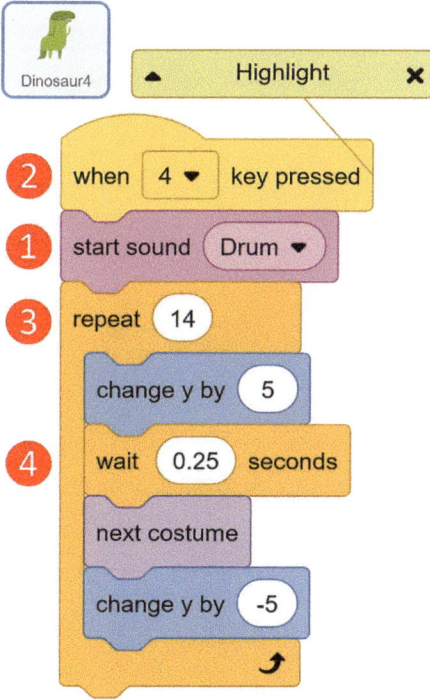

Now we've got all four dinosaurs able and ready to jump in with their own tracks. Give it a test run to make sure each dinosaur responds to the correct key, plays the correct music, and animates while their music plays. Does *Dinosaur2* seem off?

If you're looking to learn more about animation in Scratch, our Interactive Story project in Book 2: Intermediate has some great techniques to know!

Step 9: Double Up

You'll probably notice that *Dinosaur2*'s music is much shorter than the others and its timing doesn't quite work. We could drop its ●[**Repeat (#)**] and ●[**Wait (#) Seconds**] times to adjust, making it dance less, or dance quicker, but here's a nifty trick to try instead. If *Dinosaur2*'s music is half as long, what if we make it play it twice?

You might think that means copying and pasting what we've got, and that would work, but good coding is, in some ways, being lazy. What is the simplest and easiest way we can do that right? Well, we've already discovered

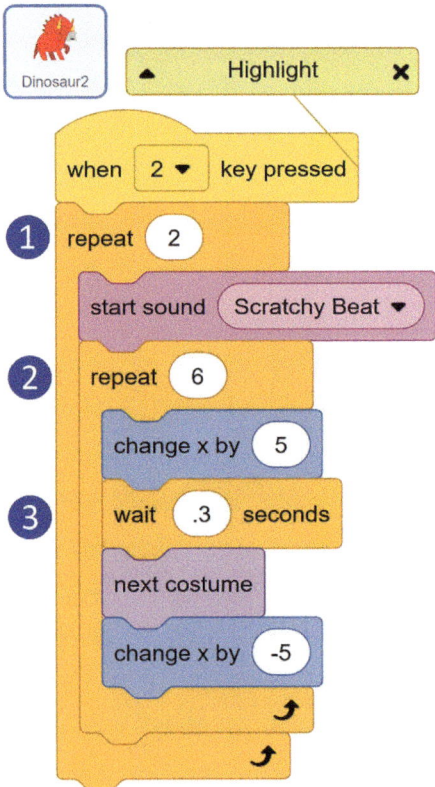

the ① •[Repeat (#)] code block. So let's use it to solve our problem. Go to the orange •Control code blocks and grab another •[Repeat (#)] code block. This time we're going to put it directly below the •[When (2 Key) Pressed] and above the •[Start Sound (*sound*)] code block. All the code blocks below should automatically move inside it, but if they don't, simply move the start sound and the other repeat inside it. Now, if we change the outer/first •[Repeat (#)] to 2, it'll play the music twice, doing the dance routine while it plays both times. The inner repeat runs until it finishes before the outer loop can repeat itself, which will run the inner loop all over again! We just need to tweak the ② inner/second •[Repeat (#)] to 6 and the ③ •[Wait (#) Seconds] to 0.3 seconds. Now it should be a great match of music, dancing, and have similar timing to the other dinosaurs.

If you're interested in more music-themed projects, we've got another in Book 3: Advanced in the Point-and-Click adventure game.

Step 10: Volume Control

The last thing we'll do for this project is add in some volume control for the background music. Five different songs playing at the same time is a bit much, so let's give players the ability to turn the background music on or off. We'll do this by clicking on the word *"Stage"* on the right side of the screen again.

Here we see our background music code. We need to add a few more code blocks so we can basically mute the music on command, if desired. For this, let's go to the yellow •Events code blocks and grab two •[When [*Space*] Key Pressed] code blocks. One we'll make into a ① •[When [Up Arrow Key] Pressed], and the other into a ② •[When [Down Arrow Key] Pressed]. The

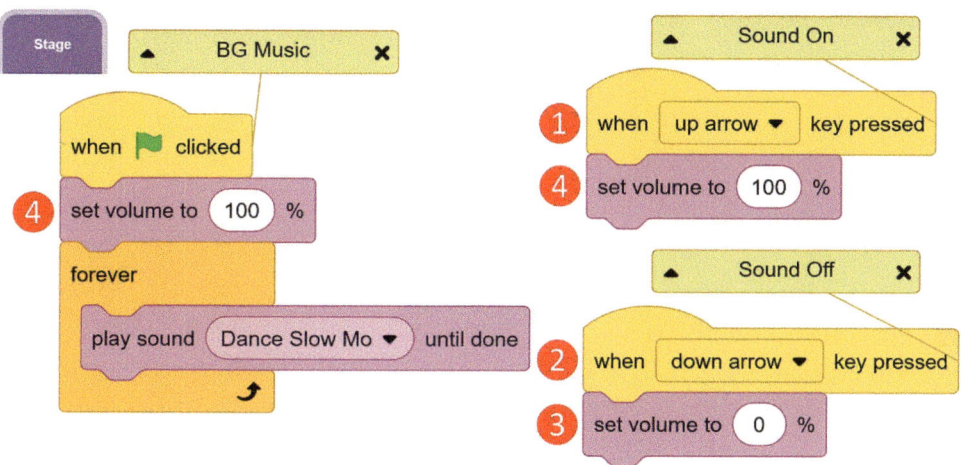

player will press up to raise the volume/unmute in *//Sound On*, and the down arrow to lower the volume/mute in *//Sound Off*.

We'll need three ●[**Set Volume to (100%)**] code blocks. As a reminder, they are found in the pink ●**Sound** category. One of each of these code blocks will go under one of the ●[**When . . .**] code blocks already in place. Since we want pressing down the arrow to mute the background music, set the percentage of ③ ●[**Set Volume to (%)**] to 0% on the code block underneath ●[**When** [*Arrow Down*] **key pressed**]. The other two ④ should remain at 100% since they are setting the initial volume in the ●[**When** ▷ **Clicked**] event, as well as raising it in the case of the user pressing the Up key.

The player can now mix and match the dinosaurs' songs with or without the background music and see them dance along to the music. This finishes our first project. You're now a Scratch coder!

For working copies of this and every project in the book series, visit www. massivelearning.net *for direct links to Scratch projects and to see our other projects and resources for coding education!*

7

Beginner Project 2: Fireworks Display

What This Project Is

Our second project focuses on the artistic side of Scratch. We'll be creating a fireworks display, letting the player time the launches of fireworks to their arrangement. Here we'll be expanding on our knowledge of Scratch by looking into the Costumes tab more while reinforcing our earlier work with animation, using more motion and graphical effects. It will take less than an hour for beginners to find their way through, but the design is open-ended to make as complicated a firework show as the user wants, using the same fundamental techniques.

What We're Learning with It

In this project, we turn our attention to graphics. We'll be focused on animations, art assets, and graphical effects. This project is also more open-ended as a creative tool for students to explore, allowing them to plan out and design their own ideas.

Sprite Properties. We'll learn about reading, changing, and resetting sprite properties, a key to understanding object-oriented programming.
Resets. The importance of reset functions for project development and replayability.

DOI: 10.4324/9781003399018-7

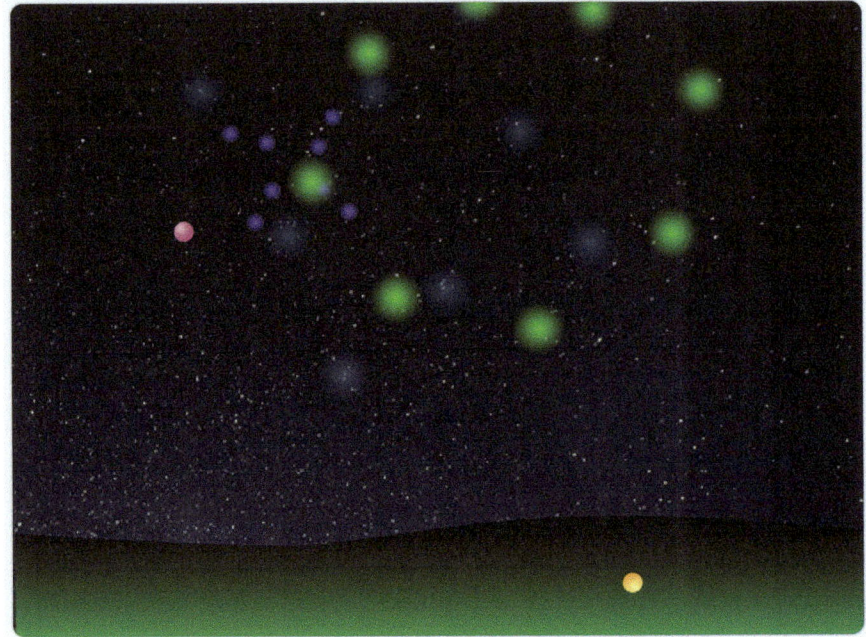

Figure 7.1 The Fireworks Display project finished.

Creating Art Assets. Get hands-on in creating (simple) custom art for our own sprites.

Visibility. Turning sprites visible and invisible.

Graphic Effects. Using the built-in graphic effects to alter the appearance of sprites.

Sizing. We'll work with dynamically sizing sprites with code.

Concurrent Events. How to use events to trigger simultaneous action across multiple sprites.

Working with Gradients. We'll use more of the graphics tools and incorporate gradients into our designs.

Building It

Step 0: Create Your New Project

Make sure you're logged in to Scratch, then click **Create** to begin a new project! Since we won't be using it, we can delete the **Scratch Cat** sprite by clicking on the trash bin on that sprite's thumbnail in the Sprite Listing.

Step 1: Creating the Background

We'll start by setting the scene for our fireworks show. A nice dark night sky will help make our fireworks easily visible no matter their colour. We'll use a built-in option, but alter it to our needs. Click on the **Choose a Backdrop** button in the bottom right corner of the screen. Select the **Stars** background to become the active **backdrop**. Now we're going to edit it.

For this step, we'll need to switch to the **Backdrops** tab, which is the **stage**'s equivalent of **sprite**'s **Costumes** tab. This is where we can make and edit art in Scratch. With the **Backdrops** tab selected, you'll see the code space has been switched to a basic digital art program. Our selected **Backdrop** should be visible, and both above and to the left are various art tools to work with. Let's try drawing a foreground, a small strip of dark land that our fireworks will start out on. But before we can edit it, we need to click the button below the image that says **"Convert to Vector"** so we can use the vector drawing tools.

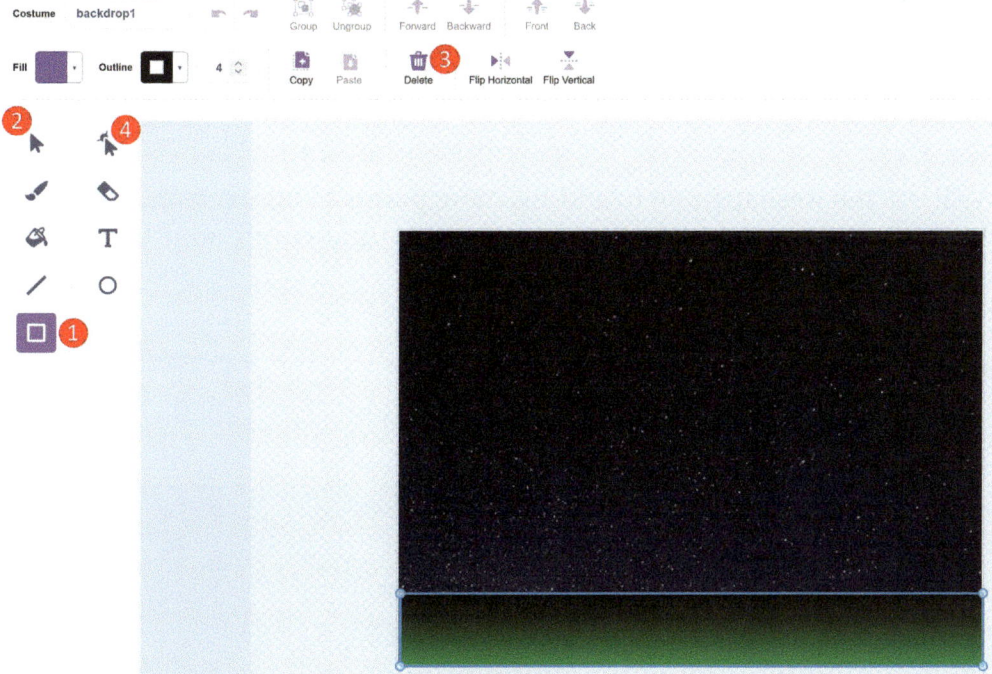

We'll start with the ❶ **Rectangle tool**. Click on the little square icon in the toolbar on the left to select the **Rectangle tool**. We need to draw a rectangle the full width of the image (or a little more) along the bottom portion of the image. Simply click where you want it to start, and drag to the opposite corner and let go to create a rectangle. If you don't like what you make, you can select the ② **Select tool** (arrow) on the left to select the rectangle and either resize it by clicking and dragging the corners, or you can just delete it with the Delete key or the ③ **Delete** trash bin icon above the drawing, but be sure to select it first!

After creating a rectangle the right size and position, we can then make a few tweaks to make it look appropriate. First, we'll need the ④ **Reshape tool** on the left. It looks like an arrow on top of a dot with two lines coming out of it. This is a very powerful and useful tool, but it can be a bit tricky to get used to, so here I'll introduce it. You'll want to practice with it to really get the hang of it.

Shapes

*Shapes in Scratch are made up of points. The computer draws lines between these points to create your shapes. The points can connect to other points using straight or curved lines, and they can have different rotational properties to change how curved lines will connect. The area between all the lines is coloured with the **Fill** property; the lines are coloured with the **Outline** property and thickness. Lines can be drawn or not, and the area can be filled or not. There are lots of options, but the key is to explore and experiment to get used to the tools and figure out how to use them best.*

We'll use the **Reshape tool** to add two new **points** to our rectangle. We want the top edge of our rectangle to be a nice, natural curve, so it looks like a

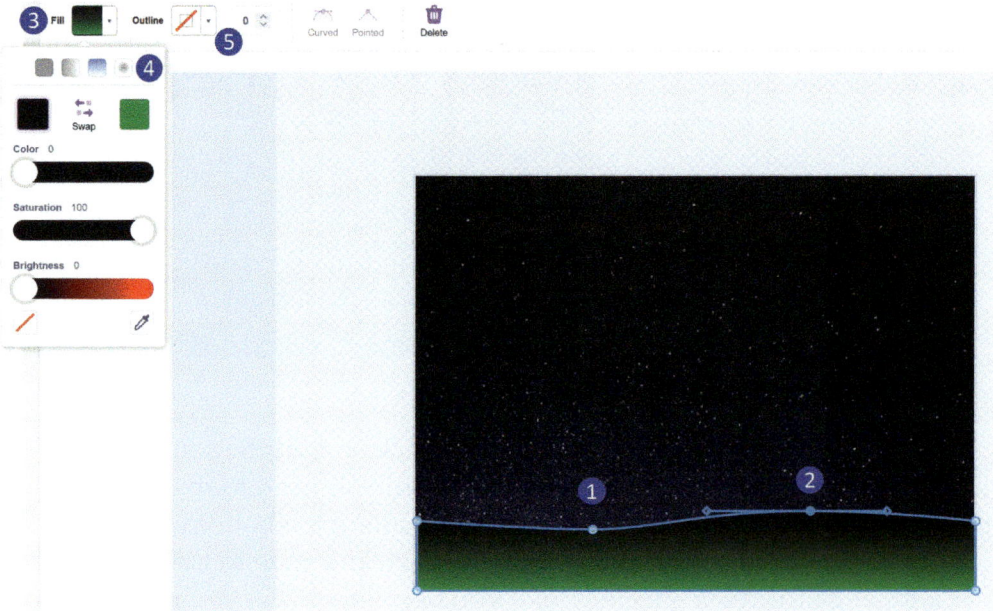

horizon. We can click along the top line of the rectangle about ❶ 1/3 and ❷ 2/3s along its length to add new **points**. After creating the new **points,** click and drag them a little off the main line; try to drag one above and one below the original position. You should be left with a nice serpentine curve.

Next, we need to change the colouring of our shape. To the top and left of the drawing, you'll see a ❸ small square labelled **Fill**. Clicking on this will allow us to change the colour of our selected rectangle. You can use the sliders to create any **colour** you want by selecting a **colour** (or hue), **saturation**, and **brightness**. However, we're going to use a special feature that's easy to miss. Above the colour selector, there are ❹ four icons showing different colour patterns: **solid**, **left/right gradient**, **up/down gradient**, and **radial gradient**. Gradients allow us to use two **colours** that will be smoothly blended from one to the other in a specific pattern. Click on the third icon for the **up/down gradient**. You'll now see you get to select two **colours**. Make one colour black (**brightness** 0), and the other a dark green. Make sure they're in order so that the green is on the bottom and the black is on the top of our shape. This makes it blend from light at the bottom (closest to the viewer) to black at the top (furthest from the viewer) to give a bit of perspective to our scene, and allow it to blend nicely into the night sky.

Lastly, we'll turn off the line drawing for the image by clicking on the ❺ little square labelled **Outline**. At the bottom left of the colour selection panel, you'll see a square with a red diagonal line through it. This option indicates to not draw or be transparent. If we click on it, there will no longer be a line drawn around the edge of our shape, making it look more natural. We now have the perfect **backdrop** for our fireworks display!

Step 2: Creating Firework Sprites

With our **backdrop** art done, we can turn to our fireworks now. We'll again use a hybrid of built-in and custom art for our fireworks. We'll start by adding a **sprite**. Click on the **Choose a Sprite** icon on the bottom right-hand corner of the screen, the one that looks like a cat face. You'll see there's no firework sprite built in. Instead, we're just going to use the *Ball* sprite as a very simple base, so click on it.

Now that we've got our *Ball* added as a **sprite**, let's switch over to the **Costumes** tab and customize it. We can see it has five **costumes**, each a different **colour** by default. This way, we can select a default **colour** for it to start as. This *Ball* image will be the unexploded firework that will first sit on the ground, then launch up into the sky. Now, we need to add in the exploded version of our firework. Thankfully, we can create a very simple version to start with.

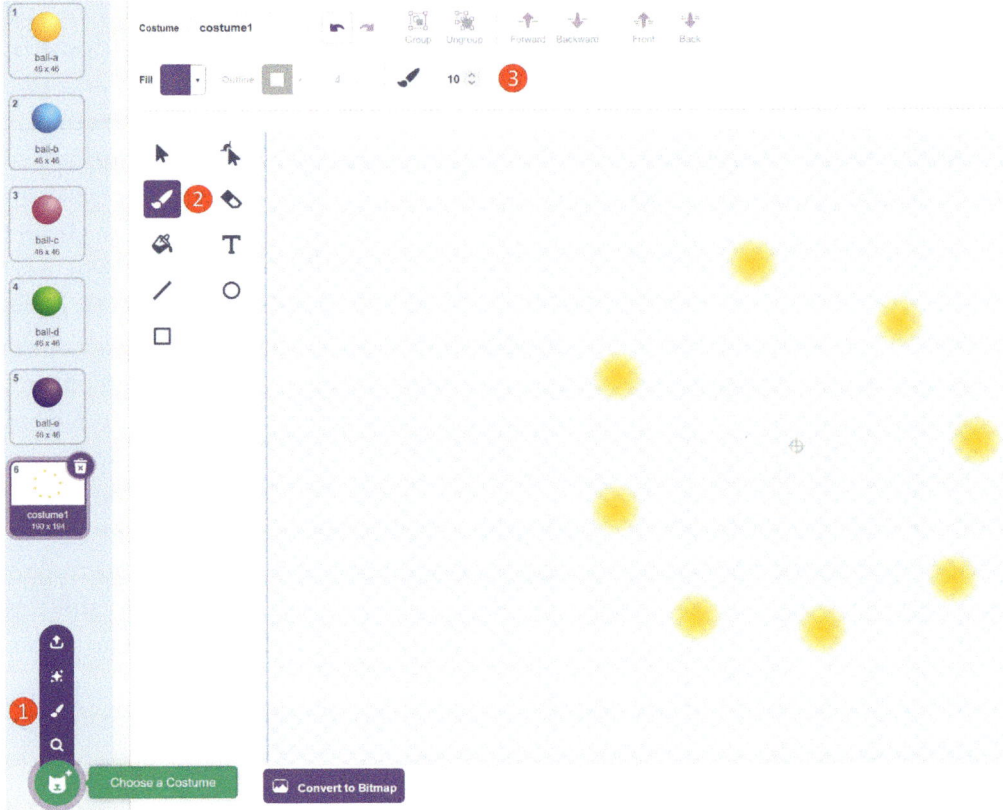

To begin, we'll create a new blank **costume**. To do that, hover your mouse over the **Choose a Costume** button in the bottom left-hand corner of your screen. You'll see a little bar extend out of the button. We want to scroll up to the ❶ little paintbrush icon on that bar labelled **Paint**. When you click on that, it will add a new, completely blank **costume** to the **sprite** for us to add new art to.

Now let's draw our exploded firework. To keep things simple, we'll just use the ❷ **Brush tool**. Click on the paintbrush icon on the left-hand side to select it. Make sure we've got the right view of our **canvas** (drawing area). In the bottom right-hand corner of the **canvas**, you'll see some zoom icons: −, =, +. To zoom in, click the plus; to zoom out, click the minus; and to reset to full view, click the equals sign. Click the equals sign to make sure we've got the normal-sized view of our art.

To create our exploded firework, we need to set the properties of our brush. Above the drawing, once the **Brush tool** is selected, you will see a ❸ **Brush Size** property. Click on the number and change it to 50. This determines how big a circle your paintbrush makes when it is clicked or how wide the stroke is when you move it. We won't worry about the **colour** we're painting with for now.

Now we need to make our exploded firework. Imagine how it should look when it explodes. A cloud of dots of **colour**. Try making a big circle pattern out of small dots to start with. Once we get the firework working, you'll see how this looks and be able to make other types of firework, understanding how it will be used, so keep it simple for now.

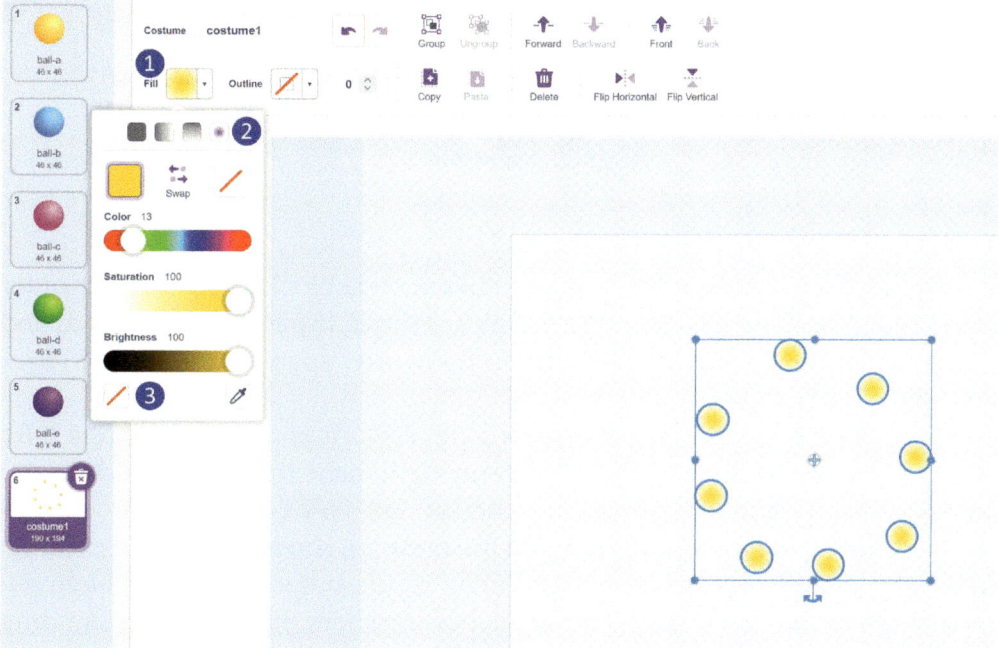

Next, we'll change the **colour** of our firework. Choose the **Select tool** (the arrow) on the left-hand side. Then click and drag to select all the dots you just drew. We can change all their **colours** by clicking on the ❶ **Fill** box to the top left of the drawing. Just like the foreground, let's use a gradient to make our firework look really nice. Select the fourth pattern icon, the ❷ **radial gradient**. Choose a **colour** you like for the centre colour, and then select the ❸ bottom left box with a red diagonal line through it to make the second colour transparent. This will give a nice glowing effect to how the firework is drawn. Click on one of the *Ball* costumes to turn it back into its unexploded version for now. You can drag the firework to any position on the ground you'd like. Now we'll focus on the code.

Step 3: Into the Sky

For coding our firework, we want to have it launch into the sky when the player presses a button. Let's start with a simple version using code blocks we've already used in *Project 1: Dino Dance Party*.

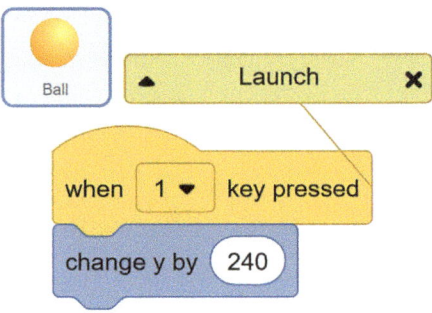

We start, as always, with a yellow ●**Event** code block to tell the computer when the code should trigger. We'll grab a ●**[When [Space] Key Pressed]** and change it to the "1" key by clicking on the word "space" on the code block and selecting 1 in the list.

Next, we'll go to our blue ●**Motion** code blocks and grab a ●**[Change Y By (#)]** code block and attach it to our ●**Event**. We'll make the change fairly big to have the firework shoot high up into the sky. Let's try 240.

Now, test out the code. Press 1 and the firework should move to a position in the sky. Not ideal, is it? It's the right end result, but let's work on having it move up into the sky instead of just teleporting there. For now, you can drag your firework back to the ground.

We're just getting started with movement and physics in this project, but by the time you get done with Book 3: Advanced, you'll be able to create your own platforming games!

Step 4: Movement Over Time

The key here is, our code made the change in one single step. In one frame it's on the ground, the next at its destination. What if we could get it to move over time, just a little bit each frame? Learning to work with the concept of frames or steps is the key to getting good animation and game flow. We can have this change happen over many frames of animation so we can see it moving from the ground to the air. Let's add a ●**[Repeat (#)]** code block to make this possible.

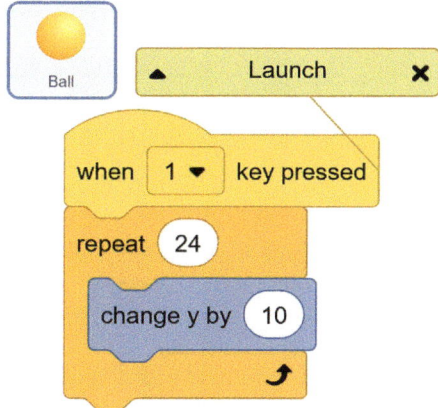

Go to the orange ●**Control** code blocks, grab a ●**[Repeat (#)]** code block, and place it under the ●**[When [1] Key Pressed]** with the ●**[Change Y By (#)]** code block inside it. Now, we don't want to move 240 ten times, so let's adjust the numbers. If 240 is the total change we want, let's make our ●**[Repeat (24)]** and our ●**[Change Y By (10)]**, so we keep the same total.

Try it again by pressing 1. Much better, right? You can explore how frames, or steps, work by trying different numbers in the ●**[Repeat (#)]** and ●**[Change Y By (#)]** code blocks. This is a good way to explore equivalent product multiplication, but one needs to keep in mind the ●**[Repeat (#)]** also as a measure of time – one repeat per frame (generally 1/30th of a second); fewer repeats mean less time for shorter and faster animations, while more repeats mean longer time frames and slower animations.

Step 5: Arcing Motion

So now we know how to animate our firework up, but what about falling? After a firework explodes, it slowly floats back down to earth as it burns up. While we're here, let's use the same technique to add a little bit of a float-down motion after it peaks. Here, we'll learn another useful method for copying code blocks.

To move down, we'll use the same technique, a ●**[Repeat (#)]** and ●**[Change Y By (#)]** code block, so let's try copying them to save time! Try right-clicking on the ●**[Repeat (#)]** code block. You'll see a little menu appear where you can select **Duplicate**. When you click **Duplicate**, you'll get a copy of that code block and anything attached in or underneath it, so you should get a ●**[Change Y By (#)]** code block inside a ●**[Repeat (#)]** block. Simply drag these below the first set. We'll change the **①** ●**[Repeat (#)]** to 10 and the **②** ●**[Change Y By (#)]** to -3. This will move our firework down, but slowly and just a little. It's a minor effect, but it'll look great in our final product.

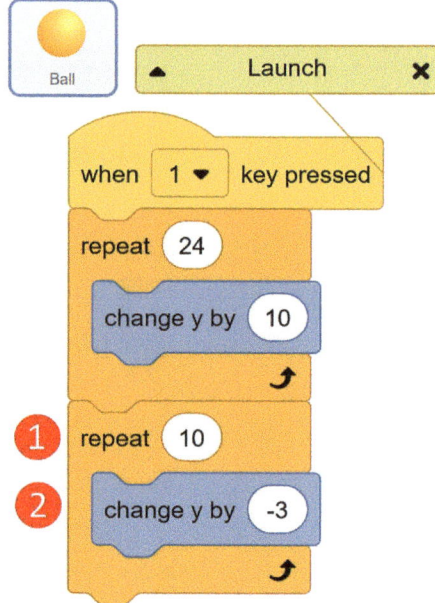

If you're interested in learning more about physics modelling and more complex motion, you can check out our Snowball Fight project in Book 2: Intermediate.

Step 6: Reset

You're probably frustrated with having to drag your firework back to the ground every time you test it. In this step, we'll make a Reset button to do the work for us. Resets are a very handy and important tool for coding, especially for teaching and experimenting. Making mistakes is part of the learning process, so we need to expect things to not work on their first try. If we can make a reset that puts everything back to exactly how it started, we can accurately compare methods and results as we build or adapt our code. Having a reset doesn't just make our lives easier; it makes it easier to understand our code and the changes we're making. So most projects should start by making a reset function. It's just a good habit.

Resets

Resets are important for another reason – we need the user to be able to use the project. Users and creators experience Scratch differently. On a project page, you should be able to play or experience the project successfully without having to see inside and edit the code. We need to make button presses or other triggers available to the user of a project. It's important to remember that when we're designing, we might be able to click and drag objects where we like, but without the ● [Set Drag Mode [Draggable]] code block run for the object, users won't be able to. As a creator, we can run code in the editor by clicking on it, but the user can't do that. Reset buttons or methods should be available to our users to ensure they can experience and enjoy the project as we intended, so don't overlook making these accommodations.

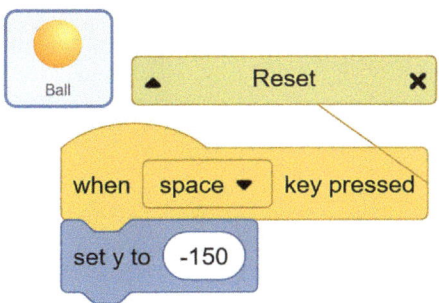

To make our *//Reset* stack, we, of course, start with a yellow ●**Event** code block. We'll grab a ●**[When [Space] Key Pressed]** to be our reset trigger. When the player hits the space bar, things will go back to where they started. To do that, we'll need a blue ●**Motion** block. Let's use a ●**[Set Y To (#)]** code block, and we'll set the value to -150. This will put our firework down on the ground, setting only its y position, since we're only changing its y position.

Try it out by launching your firework with the 1 key and resetting it with the space key.

Step 7: Explosions!

Now we get to what everybody has been waiting for – let's explode the firework! Here we'll need some purple ●**Looks** code blocks. We're going to need a ❶ ●**[Switch Costume To [costume]]** code block between our two ●**[Repeat (#)]**. This way, our firework shoots up, explodes, then falls back down. The ●**[Switch Costume To [costume]]** code block needs to have the correct costume selected. Choose "*costume1*" from the list.

Now test it out. Press 1 to launch your firework. Reset using space. Now wait . . . that's not right. Resetting isn't affecting the costume, so we'll need another ❷ ●**[Switch Costume To [costume]]** in our *//Reset* code. Add it and select one of the *Ball* choices, preferably a matching or similar colour. Now, when resetting the firework, it goes back to its unexploded ball form.

Whatever changes you make in the program, remember to unmake them with your *//Reset* function.

Step 8: Bigger Explosions!

Our costume change is great for switching to our fireworks design pattern when it explodes, but doesn't exploding generally involve changing size? Let's make our firework really explode with some size changes.

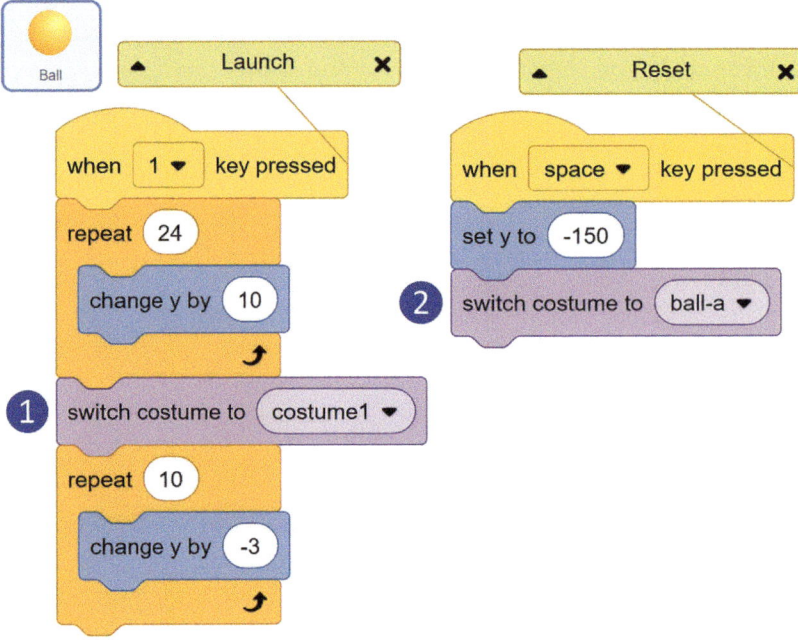

First, we'll need a new ①●**[Repeat (#)]** from the orange ●**Control** code blocks. We're going to place it between our switch costume and our falling ●**[Repeat (#)]**. Make sure the falling ●**[Repeat (#)]** is below it, not inside it. Now we'll go back to the purple ●**Looks** code blocks.

You'll see there's a ②●**[Change Size By (#)]** code block. Let's add one in our new ●**[Repeat (#)]**. This will change the size of our firework by a percentage of its original **size**. ●**[Repeat (10) {●[Change Size By (10)]}** will, in ten steps, double the **size** (10 × +10% = +100%). But remember: if we're changing something, we need to unchange it in our Reset part of the code. So let's get a ③●**[Set Size To (#)]** code block and have it in our Reset. Set the value to 25%, so our unexploded firework is much smaller than it has been – it was a bit oversized, right?

Try it out now. Looks pretty good, huh? Our explosion now actually explodes! Let's add one more tweak while we're at it. Instead of it exploding the exact same way every time, how about we randomize its rotation, so each time it runs, it'll be a little different? Go to the blue ●**Motion** blocks. Here we can grab a ④●**[Point In Direction (#)]** code block and add it to our Reset. With our firework being round and small, no one will notice if it just rotates at the start and keeps that new rotation right from the start. But here we just have a single set value for the rotation direction. We need to use a new code block if we want to randomize it.

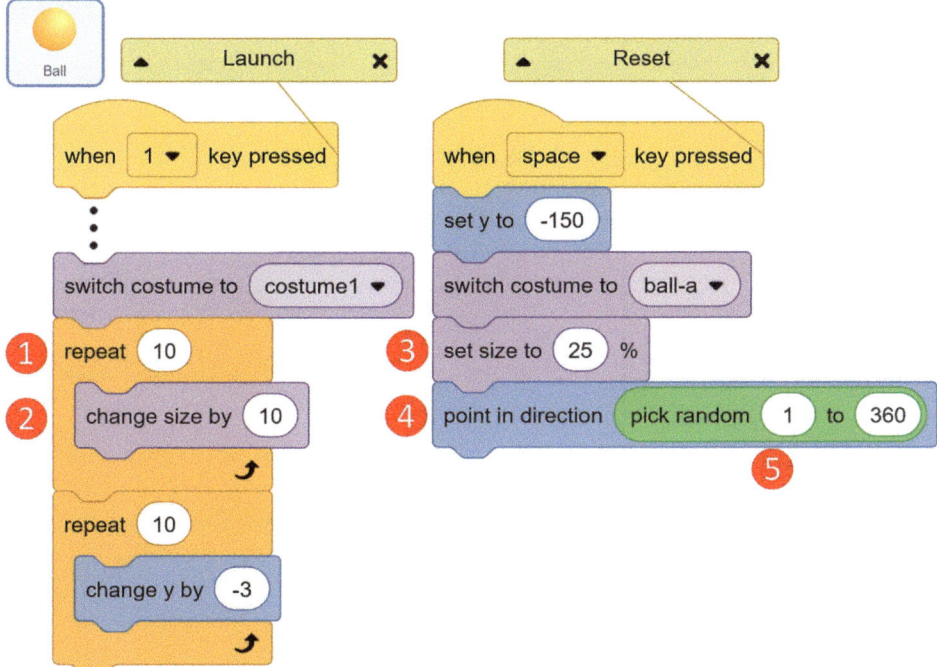

Take a look at the green ●**Operators** code blocks. You'll see a round-edged code block ⑤ ●**(Pick Random (#) To (#))**. We're going to put it in the value of the ●**[Point In Direction (#)]** code block – round edges for a round hole. With the ●**(Pick Random (#) To (#))** in place, we just need to change the values from 1 to 360 so it can be any random direction.

It's a subtle little tweak, but randomizing functions like this can be a really nice way to add some surprise, complexity, and variation to projects.

Step 9: Visibility

Right now, the biggest annoyance with our project is that we get our firework stuck in the sky, just hanging there, visible for eternity. How can we make our firework actually burn out and disappear? Let's work with the **Visibility** property of objects.

You might have noticed in the Properties panel that all objects have a **visibility** status – either visible or invisible. While we can set this in the **Properties panel**, it can also be done in code. Let's go to the purple ●**Looks** code blocks. Here we can find two very small code blocks – ●**[Show]** and ●**[Hide]**. These turn objects visible and invisible, respectively. Let's have a ❶ ●**[Hide]** code block at the end of our //*Launch* sequence and a ❷ ●**[Show]** code block in our //*Reset*. If you use ●**[Hide]** and you don't use ●**[Show]** in your reset,

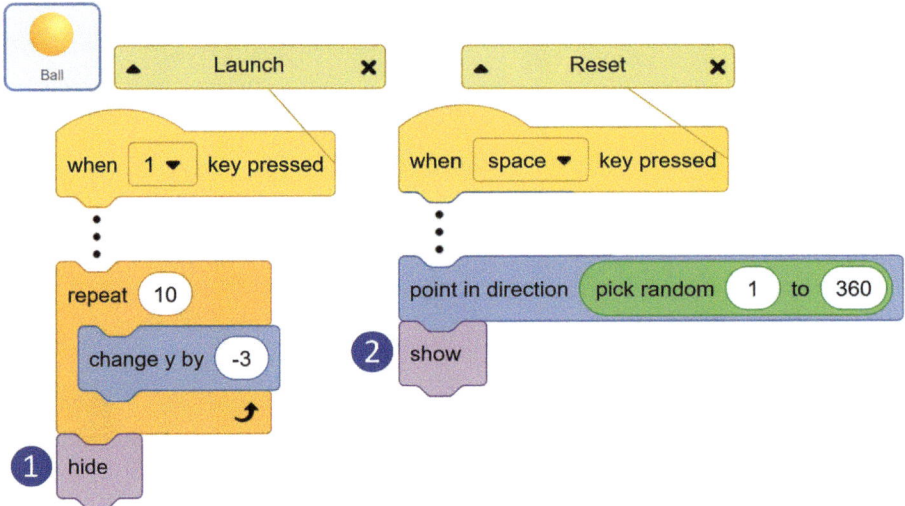

your object won't turn visible again! This is one of those things I always forget to do until I test my project and see it mess up. Testing often helps catch these mistakes.

Now our firework properly disappears after exploding but reappears unexploded when we reset.

Step 10: Sound Effects

Part of the effect of fireworks is the sound. Let's add a quick little sound effect to our project to highlight the explosion. By switching to the **Sounds** tab above the code block list, we can add a new sound effect. You'll see that there's already a Pop sound effect attached to the sprite by default. For this project we can use this sound effect. If you wanted to choose a different one, you'd click on the **Choose a Sound** button in the bottom left-hand corner, scroll down the list, and select the sound effect you'd want. Every sprite in the Sprite Library has at least one automatically loading sound. The *Ball* always has both **Boing** and **Pop** loaded, so you can just use those without loading a new sound. Whatever your choice, now we can go back to the **Code** tab to call the sound effect.

In the pink ●**Sound** code, blocks grab the ❶ ●[Start Sound (sound)] code block and place it directly beneath the ●[Switch Costume To (costume1)] block. This will play the sound effect when the costume changes and let the animation sequence proceed as it plays. A nice little audio cue perfectly timed in our sequence.

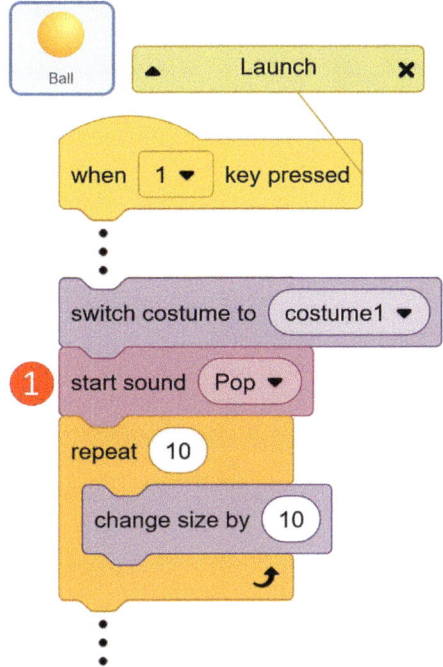

Step 11: Transparency Using Ghost FX

While we've used a lot of great techniques to get a good animation of a firework already, let's add one more technique to explore the graphic effects features of Scratch. We'll need to go to the purple •**Looks** code blocks for this.

Grab a ①•**[Change (Colour) Effect By (#)]** code block and add it to the *//Launch* stack in the falling •**[Repeat (#)]** loop just above the •**[Change Y By (-3)]** code block. If you click on Colour, you'll see a number of different options. The *ghost* effect is the transparency of an object. At *ghost* 0, an object is completely opaque (it has no ghosting), and at *ghost* 100, it is completely transparent or invisible. Using dynamic transparency with the *ghost* effect can make for some really great effects, from wispy clouds to fading fireworks. Change the value in the •**[Change (Ghost) Effect By (#)]** block to 10. With ten repetitions, this will take our firework from opaque to invisible, letting it fade away into the night.

Now, importantly, if we've ghost effect changed an object, we need to make sure our Reset takes this into account. Here we can use one of two choices. We can use a ②•**[Set (Ghost) Effect To (0)]** or a ②•**[Clear Graphic Effects]** code block in our Reset. Either will work. If we were using multiple

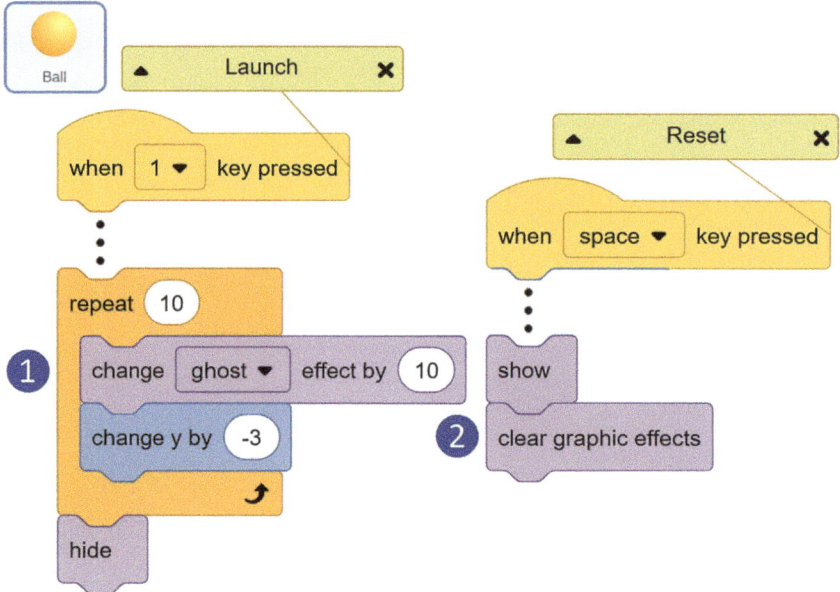

effects, a •[Clear Graphic Effects] code block would reset all of them (colour, ghost, whirl, etc.) in one go, which you may or may not want.

With that, we should have a fully functional firework for our project!

If you really want to take your fireworks show to the next level, you can learn about procedural generation in our Scrolling Shooter game in Book 3: Advanced.

Step 12: Duplicating More Fireworks

With our firework complete, the only thing our fireworks show needs are more fireworks. If you right-click on the thumbnail of the firework in the Sprite Listing, you can choose to **duplicate** the object. This will allow you to make copies of your firework, so you can easily make many more fireworks that function the same way.

With your copies, you can go into the **Costumes** tab to change their ❶ colours or explosion costumes. In the code, you can change how high they shoot by changing their •[Repeat (#)] and •[Change Y By (#)] values. Changing the ❷ key press will either launch them on the same or different key press events. You could also add in some •[Wait (#) Seconds] code blocks to time them. Now that you know the methods, adapt them to make it your own!

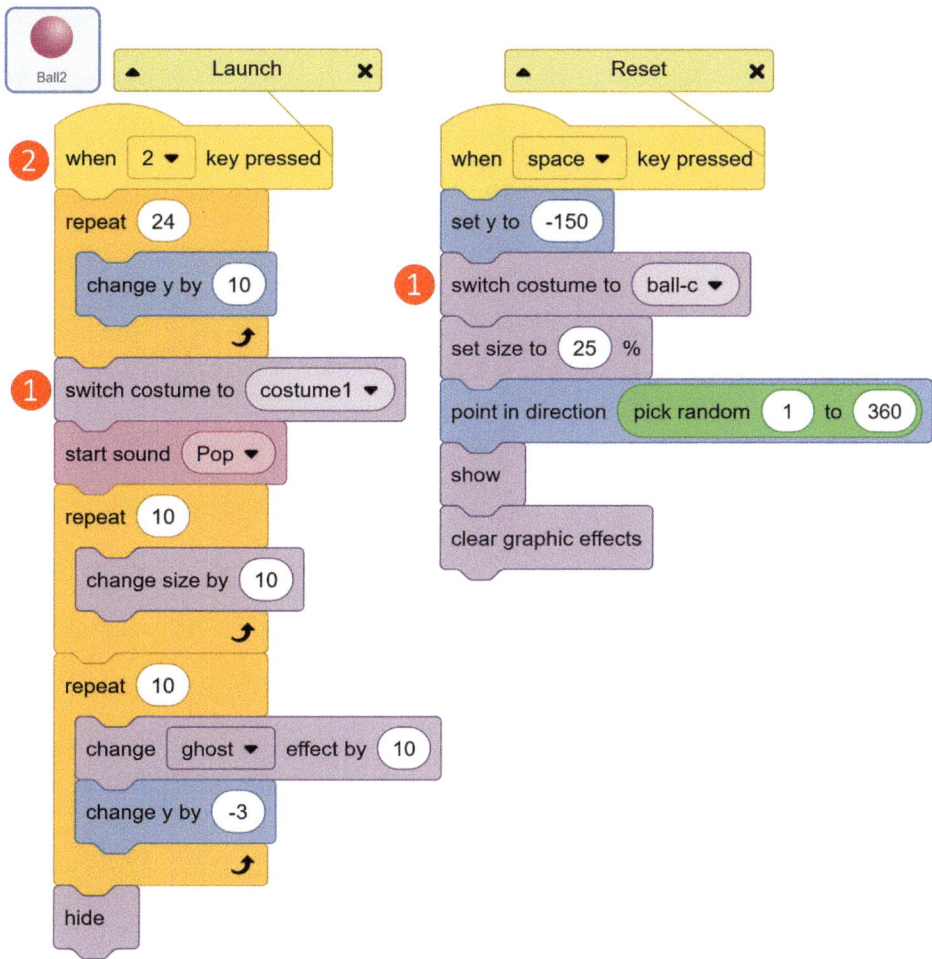

For working copies of this and every project in the book series, visit www. massivelearning.net *for direct links to Scratch projects, and to see our other projects and resources for coding education!*

8

Beginner Project 3: Batty Flaps

What This Project Is

This project is a clone of the popular Flappy Bird game that took the world by storm in 2014. In this project, we'll learn to make a player-controlled character affected by gravity that can flap to rise up, needing to pass through randomly positioned gates to proceed and earn points. In addition to the basic gameplay programming, we'll explore some graphic creation and include a title and Game Over screen to break up the gameplay. The programming and graphics for this game should take an adult around 30 minutes to create.

What We're Learning with It

For our first game project, we're going to start with something very simple – just a one-button-press game – but we'll learn a few little tricks too. With our simple gameplay, we can introduce the concept of variables in a straightforward score-keeping system. To follow up on our graphics work from the Firework Display project, we can learn some of the nice little touches of polish for a project on Scratch. We'll add in a title screen, then transition to gameplay, and end with a Game Over screen. To manage this, we'll learn to use the custom events system to time out how we want things to proceed. This will also allow us to make a game that can be replayed without using the ▷ button,

DOI: 10.4324/9781003399018-8

Figure 8.1 The Batty Flaps project finished.

which is a nice, player-friendly consideration to include in games, which also forces us to think in new ways about how code stops or runs.

> **Variables.** In computer science terms, *variables* store data for our game to process and react to. They are also a very useful tool for demonstrating the value of learning variables in math class.
>
> **Oversized Objects.** To make our gate system to challenge the player, we'll need to make objects bigger than the screen allows.
>
> **Randomized Play.** We'll use some code to introduce random variations to switch things up.
>
> **Title Screens.** Starting our game with a title screen forces us to trigger gameplay to start without using the ⚑.
>
> **Game Over Screens.** When the player loses the game, we'll show a specific graphic to signify the change in game state.
>
> **Game Replay.** Our design will allow players to continue playing the game without restarting it using the ⚑, making for a more seamless and simple experience for players.
>
> **If Statements/Conditionals Events.** We'll get familiar with ●If statements to make conditional events.
>
> **Messaging.** To create a replayable game, we'll use the broadcast system so sprites can send and react to messages.

Building It

Step 0: Create Your New Project

Make sure you're logged in to Scratch, then click **Create** to begin a new project! Since we won't be using it, we can delete the Scratch Cat sprite by clicking on the trash bin on that sprite's thumbnail in the **Sprite Listing**. We can also start by setting our *stage* to the **Castle 2** backdrop by clicking the **Choose a Backdrop** button in the bottom right-hand corner of the Scratch editor.

Step 1: Flying Bat Character

The player will be controlling a *bat* character in this game; conveniently, there's a great *bat* sprite we can load by clicking the **Choose a Sprite** button in the bottom right-hand corner. In the Properties panel, you can shrink the *bat* by setting its **size** to 50; this let's it fit our game a little better. With that done, we're going to reorder its costumes a little so they'll make more sense. At the top-left corner of the Scratch editor, click on the Costumes tab so we

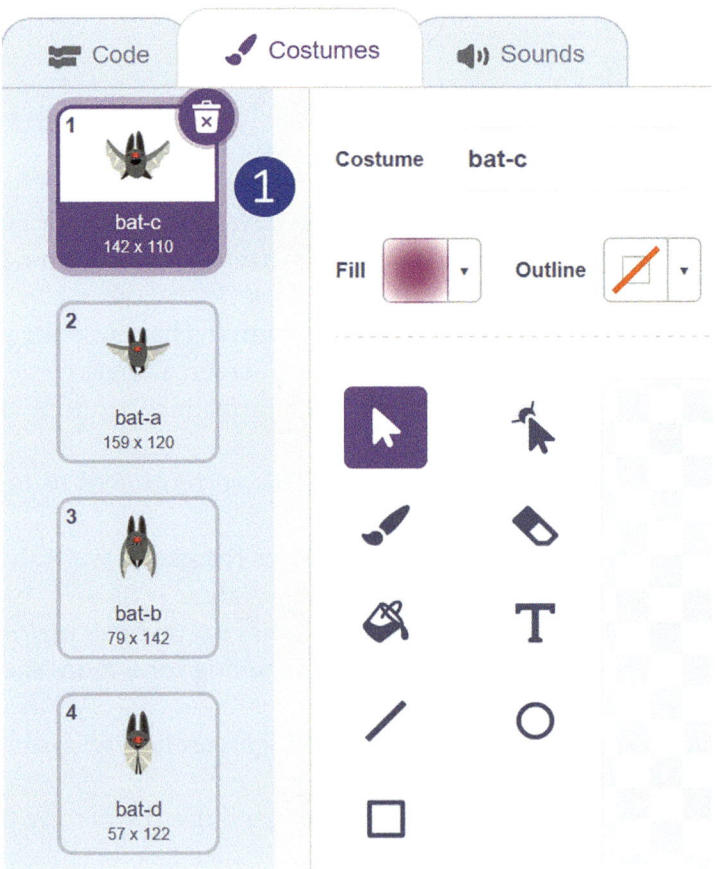

can change the *bat's* costumes. Here you'll see the sprite comes with four costumes built-in. All we want to do is ➊ reorder the Bat-C costume to the top of the list. This gives us a nice order for animations – wings up, wings out, wings down, wings closed. That's all we need to change in the costumes, so now you can click back to the Code tab so we can program some controls.

For our *bat* to fly, the player will press the Space button for one flap of its wings. This will create a short burst of upward movement. To achieve this effect, we'll need to start with a ➊ •**[When [Space] Key Pressed] Event** block to start our *//Flight* stack. Under this we'll place a •**Control** block – ➋ •**[Repeat (10)]** and set its value to 4 instead of 10. Inside we'll place a ➌ •**[Next Costume]** from the •**Looks** blocks; this will cycle through the four costumes of the *bat* sprite. Lastly, in the **Movement** code blocks, we'll grab a ➍•**[Change Y by (10)]** and change its value to 5. This way, each time the space key is pressed, the *bat* will rise up a total of 20 pixels and run through its full animation cycle.

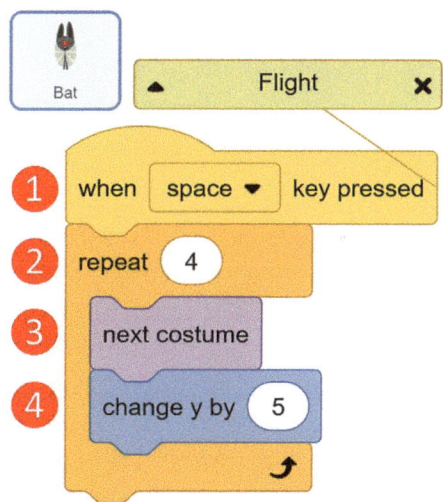

Step 2: Gravity

Our code so far lets our *bat* fly up, but there's nothing pulling them down. We're going to add in a gravity effect that will constantly pull the *bat* down, and between these two forces, the player will try to control the *bat's* altitude to avoid collisions. For our *//Gravity* effect, we're going to add in a new •**Event**: the ➊ •**[When I Receive ["*Message1*"]** code block. This is a special kind of event; instead of triggering in response to a button push, mouse click, or the game starting, it happens only when we tell it to happen. It's our own custom event system that we can use to trigger code. When you pull one out, you'll be able to click on the **["*Message1*"]** section; this will allow you to change which message it responds to. We can click "**New Message**" and make a "*Game-Start*" message that we'll use. This is the event we'll use to start playing the game, instead of just having everything run when we click the ▷. You'll see why we want this instead of the ▷ event a little later on.

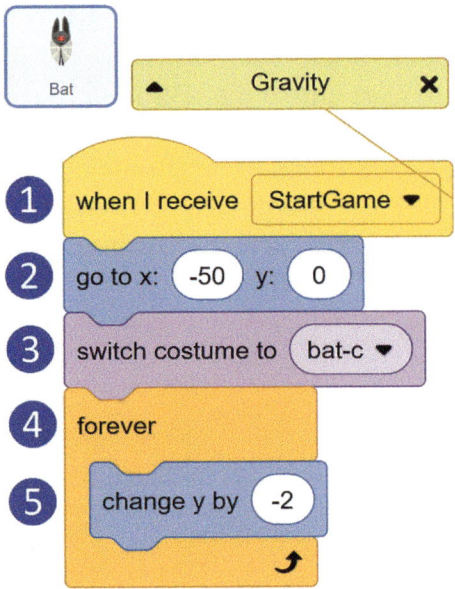

In this new •**Event**, we'll need a •**Motion** code block ②•[**Go To X:(-50) Y:(0)**] to make sure the bat is in a neutral starting position, and we'll make sure they ③•[**Switch Costume To (*Bat-C*)**] from the •**Looks** blocks. Under this we'll add in a ④•[**Forever**] loop from the •**Control** blocks so that our gravity effect will constantly affect the player. Inside this we'll place a •**Motion** block – ⑤•[**Change Y by (-2)**]. Every step of the game will slowly pull the *bat* down. To try it out, you can just click on the •[**When I Receive ["GameStart"]**] event to get it to run for now. You can see how the player will be able to flap quickly and climb, flap occasionally to stay level, or just let gravity take over to descend.

Step 3: Tree Gates

Now we can add in the challenge of the game, the gates that the player is trying to manoeuvre to pass through. We'll need to make our own graphic for this, so we need to hover the mouse over the **Choose a Sprite** button in the bottom right-hand corner until the bar of extra options pops out of the top of it. If we scroll up to the paintbrush icon and click on it, Scratch will create a sprite, but without any costumes to start with. In the Properties, we can give the sprite a name: "*Tree*".

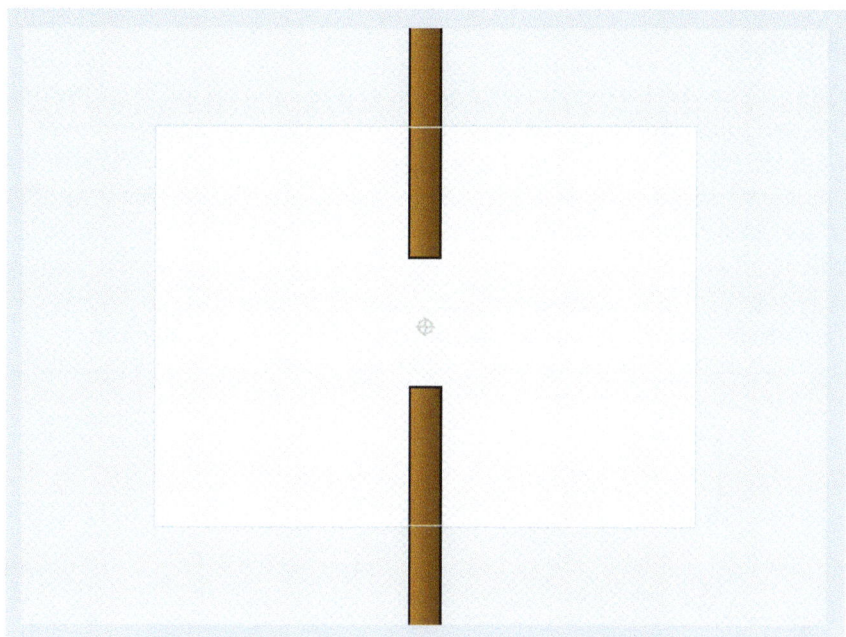

Next, we'll give it a costume, which means we'll need to go to the **Costumes** tab. Here we'll use the **Rectangle** tool, and we'll set the **fill** colour before we draw. Click on the purple Fill button. Here we'll see the three sliders that allow us to choose any colour. If we slide the **colour** to around 10, **saturation** to 100, and the **brightness** to 80, we'll get a nice light brown for our tree, but we can take a moment to make it look even better. At the top of the colour selection panel, you'll see the four fill style icons. We've used the vertical and radial gradients in our Fireworks Display project, so this time let's select the second icon – the horizontal gradient. For our tree, let's have a light brown to the left and a dark brown to the right to simulate a cylinder shape with light coming from the left.

Now that we've got the shape and colour we want, draw a thin rectangle the whole height of the workspace. Then switch to the arrow **Select** tool. You'll be able to click on the rectangle and move it; you'll want to position it so the bottom of the rectangle is a little more than one-third of the way down from the top of the screen size border and directly above the centre target reticle. Then you can copy and paste another identical tree rectangle. This one you'll want to position on the other side of the centre target reticle, about the same distance below. This will create a gap in the middle of around one-third of the screen size. That's all we need for graphics work, so return to the **Code** tab so we can get it working.

Graphics Modes

Graphics in Scratch have two modes: bitmap (or pixel) and vector. There are both obvious and subtle differences between them. In this case, we need to use vector art for our gates. Pixel art can only be 480 × 360 pixels big; it cannot extend past the edge of the screen size. With our vector-based trees, we can make them extrabig and run off the edge of the canvas. This extra size is handy because it means we can position them and we won't have the trees just suddenly end if we move the gate up or down. In Book 3, in our advanced projects, we'll learn more about just how powerful and important that extra space can be for pulling off some great features.

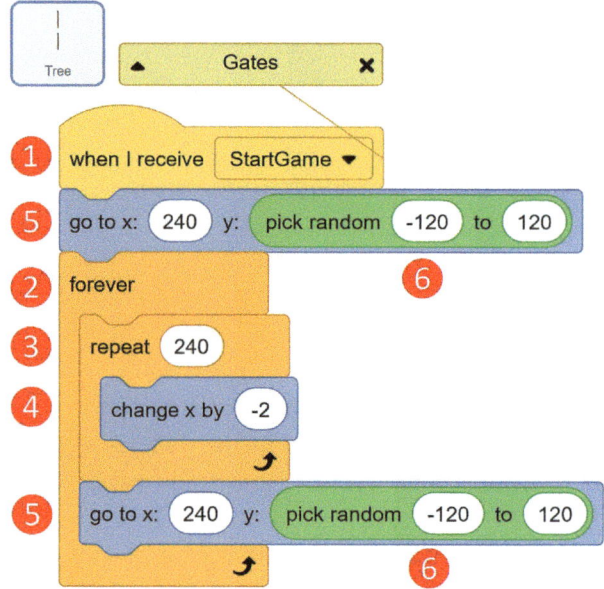

Start with another one of our new ➊ ●[**When I Receive [**"*GameStart*"]] events. Under this we'll need a ➋ ●[**Forever**] loop, and inside it a ➌ ●[**Repeat (240)**] loop both from ●**Controls**. Next, we can head to the ●**Motion** code blocks. Inside the ●**Repeat** block, we'll need a ➍ ●[**Change X by (-2)**]. Note that the ●**Repeat** amount (240) times the ●**Change X** value (-2) is -480. The width of the **Stage Window** in Scratch is 480 pixels, so these numbers combine to move the tree from the right side of the **Stage Window** to the left side perfectly. If you end up changing either number, you'll need to change the other to complete the movement cycle. Lastly, we'll need two copies of ➎ ●[**Go To X:(240) Y:(0)**], one above the ●[**Forever**] loop, and one inside the ●[**Forever**] loop under the bottom end of the ●[**Repeat**]. Next, we'll head to the ●Operators and grab two ➏ ●(**Pick Random (1) to (10)**). We'll place one in each of the ●[**Go To**] blocks in place of the y values. We'll change the numbers from -120 to 120. This will set our tree to the right side but randomize how far up or down it is, so the player will have to match altitude with a constantly changing challenge. You can see it in action by clicking the •[**When I Receive** [*"GameStart"*]] event to make it run.

> *In Book 3: Advanced, we'll teach you techniques for working with clones for our Scrolling Shooter project that show even more advanced ways to handle procedurally-generated content like we're using here.*

Step 4: Collision Testing

You'll notice that while your *bat* moves fine and the *tree* moves fine, they don't actually interact. We'll need to add in code to make those game-ending collisions happen. We'll add some code to our *bat* to make this happen. In the *//Gravity* stack under the •[**Change Y by (-2)**], we'll need to add two •**Control** code blocks: an ❶ •[**If <condition> Then**] and, inside it, a ❷ •[**Stop [All]**]. Our

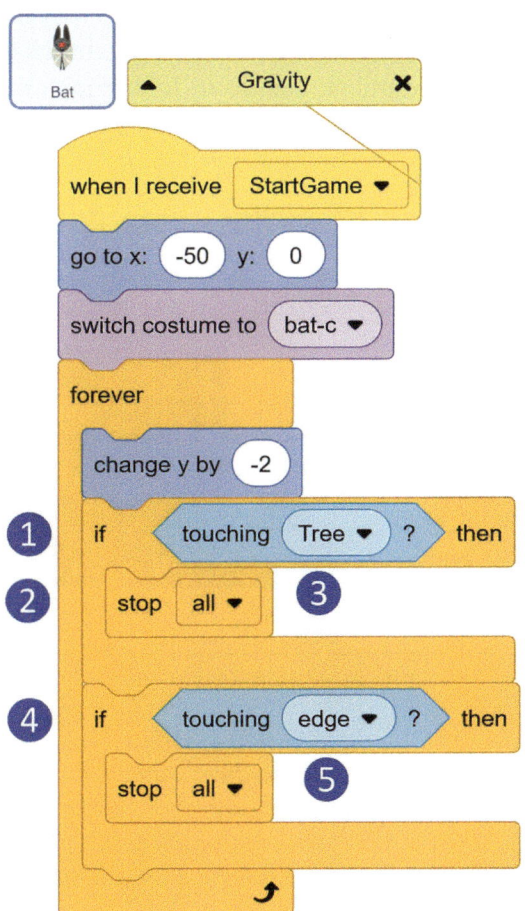

condition we'll get from the ●**Sensing** code blocks: a ③ ●**<Touching (Mouse-Pointer)>** code block. We aren't interested in the mouse pointer, though, so click on that and instead select **(Tree)**. With this ●**If** added, our *bat* will now collide with the *tree*, but we want one other risk to the player – the **edge**. If the *bat* soars too high or drops too low and touches the **edge** of the **Stage Window**, we also want the game to end. So right-click on the ●**If** and select **Duplicate**. Place this ④ duplicate ●**If** stack under the first (make sure it is inside the **Forever** loop but outside and underneath the first ●**If** and not inside it!). In the second ●**If**, change the ⑤ ●**<Touching>** block to test for the **edge**.

Step 5: Scoring

Now that we've got our game working, the question is raised: "*Well, how did I do?*" Our next challenge will be to add a scoring mechanism. Go to the ●**Variables** code blocks and click the ① ●**[Make a Variable]** button at the top of the list. We'll name ② our new ●Variable "*Score*". You don't have to change any of the other options in the pop-up, so click OK to create the variable. You'll now see a little "Score 0" info box in your game. This displays ③ the current value of the ●"*Score*" variable. In the ●Variables code blocks, you'll see a little oval ④ ●**("*Score*")**. The checkmark beside it determines whether to display the variable in-game. In this case, we want the "*score*" visible, so we'll leave it checked. With that added, we'll need two code blocks to implement the system. Go to the *Tree* code.

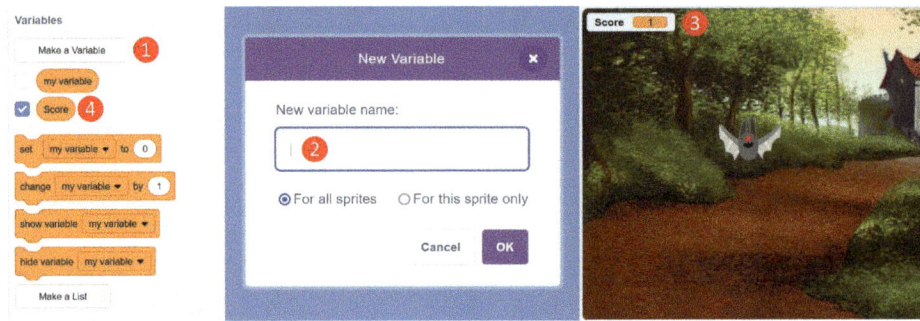

Variables

Variables are very important in coding; it's pretty hard to do much without them. I think it's important to introduce students to them a little differently than math class. In math class, variables are mostly just problems they have to deal with. Not very fun. In coding, they're a tool for them to work with. I think this can really help make for a more positive relationship. So how do we get them to understand variables not as problems but as tools they can use? Thankfully, every student you've ever taught already knows what they are and use them – they just didn't know it.

A variable is something you want the computer to remember and that can change. Anytime we need the computer to keep track of and remember something, that's a variable. When we need to have a score in a video game, that's a variable. The number of player lives? A variable. The amount of gold coins you have? You guessed it: a variable. Kids are very familiar with variables; they just didn't know the term. We can also expand this concept to the real world. How much money do you have? How tall are you? What's your address? Those are all variables. It's any data point that can change. We can interpret almost anything as variables, and that means it's a powerful tool to learn how to work with.

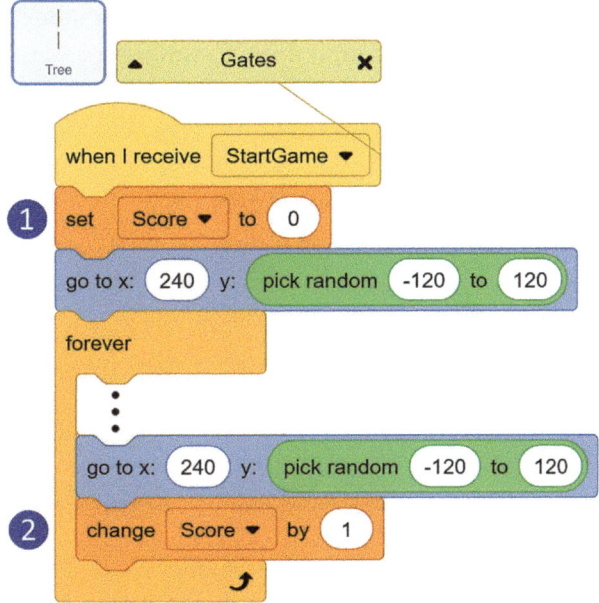

We'll add a **1** •**[Set ["*Score*"] to (0)]** to the top of the *//Gates* stack above the •**[Forever]** loop. This will ensure each new game starts from 0 points – an important but easy-to-overlook piece of code! At the bottom of the •**[Forever]** loop, under the second •**[Go To]**, we'll put a **2** •**[Change [Score] by (1)]** block – this will give the player 1 point for every gate they successfully pass when it resets position.

In the Snowball Fight project in Book 2: Intermediate, we show how to add in procedurally generated bonus pickups that could be a great addition to a project like this!

Step 6: Title Screen

The last object we'll need in our game is the *title screen*; this will function as both our **Title screen** and **Game Over** screen. Hover the mouse over the **Choose a Sprite** button in the lower right, and then click on the paintbrush icon from the options bar that extends above. We'll rename this sprite to "*Title Screen*", position it at **X: 0 Y: 0** so it's centred on the stage, and then head over to the **Costumes** tab to draw it.

For this sprite, I like to make a **rectangle** with a gradient fill, but this time try the vertical gradient (third icon in the colour selection panel) so we can have a colour starting at the bottom and blending toward the top. In this case, let's make the second colour transparent so you can see some of the game behind the *Title Screen*. To make the top colour transparent, select the first, or left, colour, then to select transparent, click on the white box with a red slash through it in the bottom left-hand corner of the **colour selection panel**. Then you can select your second, or right, colour to whatever you like. I went with orange for a Halloween theme. Draw a **rectangle** the full width of the screen area. You can choose how high you want to make it. If you cover the whole screen, it will stretch the gradient the whole way; if you make it shorter, then some portion of the screen will be completely clear.

Now that we have our background design, let's add a title for our *Title Screen*. Select the ❶ **Text** tool. Here you can select a font for our text with the font selector above the work area. You can type in some text first if you need to see it better. There isn't any centre-alignment option for text in Scratch, unfortunately, so if you want vertically aligned centred text like mine, you'll need to make each line of text its own text object separately. Here I used the ② **curly** font to be more Halloweeny and chose purple font by selecting it as the **fill** colour. To ③ resize text, you just stretch the text object to suit. If you select all your lines of text first, you can resize them all at the same time and to the same scale; just hold Shift while you click on multiple objects to select more than one simultaneously. With the text resized, I did one last thing – copied and pasted it. This gives me a second copy of the exact same text. With the copy (automatically selected when you paste), you can change the **fill** colour to black, position it offset to the original, and then click the ④ **Backward** button to bump it behind the original text to act as a drop shadow effect – perfect to make the title pop!

Switch back to the **Code** tab, and we can add the functionality. Add a ① •**[When ▷ Clicked]** event and attach a •**[Show]** block from •**Looks**. This will ensure our *Title Screen* starts our game. Then add a ② •**[When This Sprite Clicked]** event; this will allow our player to start the game. To do that, you'll need to attach a •**[Broadcast ("GameStart")]** event. This will trigger the •**[When I Receive ["***GameStart***"]]** event in every object and then a •**[Hide]**

code block from ●**Looks** to make sure the player gets a clear view of the game. You can click the ▷ to see the new system in action.

> *Wondering how you could make the background scroll along as well? In Book 3: Advanced, we'll show you how to make a scrolling background for our Scrolling Shooter project!*

Step 7: Game Over

This kind of game is endless. A player can keep playing the game for as long as they can survive; it only ends in defeat. To mark this inescapable occasion, we'll add in a Game Over screen. Select the *Title Screen* and go to the **Costumes** tab. Here we'll add a new costume by hovering the mouse over the **Choose a Costume** button in the lower left-hand corner and clicking on the paintbrush icon from the Options bar that emerges from the top of the button. This will add in a new blank **costume**. With what you learned in step 6, add a vertical gradient background with black at the bottom and transparency at the top. This will give an ominous overlay to the game when the player dies. Then to make things crystal clear, you can add in text of *"Game Over"* as well.

Switch back to the **Code** tab and we'll add in the functionality. We'll need one more ➊ ●**Event**: a ●[**When I Receive** [*"Game Over"*]] event. You'll need to click on the message and click *New Event* to add this new option. Now we'll need to add some ●**Looks** blocks. In the //*Game Over* stack, add a ➋●[**Switch Costume To (Costume2)**] and a ●[**Show**] to make sure our **Game**

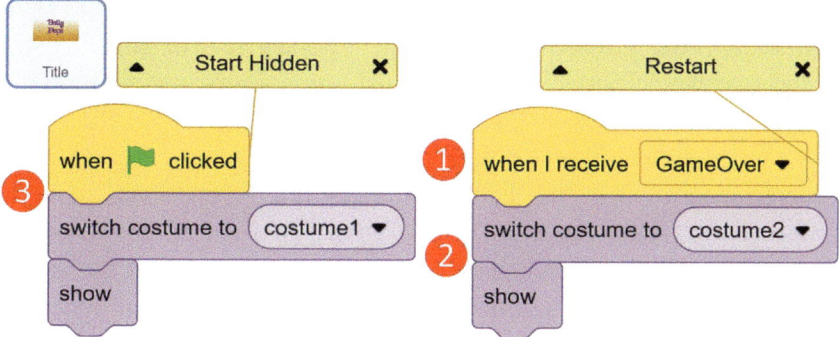

Over screen appears correctly. Now we'll also have to add in a ③ •**[Switch Costume To (costume1)]** to our •**[When ▷ Clicked]** event to make sure our *Title Screen* appears as a *Title Screen* when the game is started.

Our *bat* will need some code too. Switch to the *bat*. Now that we've created a *"Game Over"* •**Event**, we'll need to call it. In the *//Gravity* stack, you'll need to add a ① •**[Broadcast ("GameOver")]** code block above both of the •**[Stop**

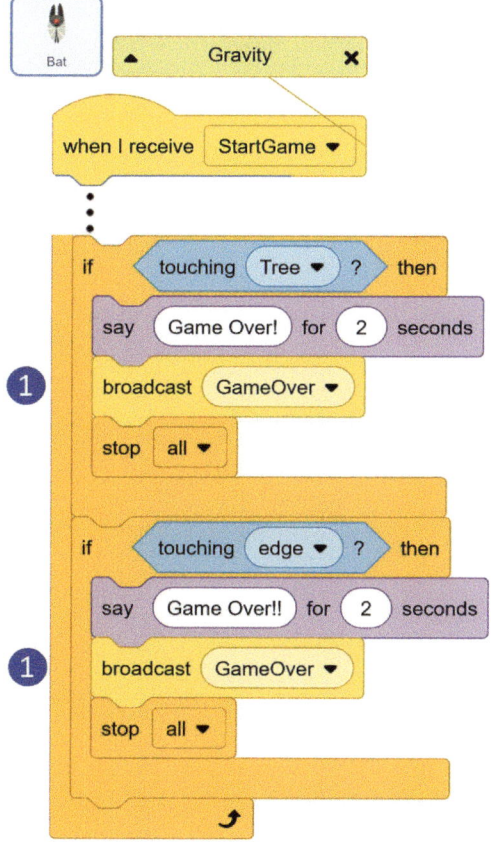

[All]] code blocks. This will ensure the Game Over screen appears when the player crashes. Now you can click the ▷ and try it out. You'll discover an unexpected feature! When you get a Game Over screen, just click on it – it will start a new game! How does this happen? Because we put our code inside *"GameStart"* •Events, when we click on the Game Over screen (which is just a costume change of the *Title Screen*), it runs the *"GameStart"* •Event. Even though we •[**Stop [All]]**, the *"GameStart"* event gets called anew, which starts that code running again! This is a benefit of what's called "encapsulating" our functionality. Everything is self-contained in custom events or scripts that we can start and stop as needed – a powerful and convenient technique.

For working copies of this and every project in the book series, visit www. massivelearning.net *for direct links to Scratch Projects, and to see our other projects and resources for coding education!*

9

Beginner Project 4: Butterfly Catching

What This Project Is

For our final project in this book, we're going to take our game-making to the next level with some more logic and controls. It will feature a butterfly flying around the screen for the player to click on. However, each time they succeed, a beetle will spawn (appear) and also run around the screen. If they click on the beetle, they'll lose a life. How long can the player succeed in a game that gets harder as they play? In around 30 minutes, for an adult, you should end up with a fun and challenging simple game great for young students while also providing training to use some of the most important techniques in coding.

What We're Learning with It

We'll learn more about conditional statements, a fundamental tool for building interactive and dynamic projects. We'll also be learning about scores and variables to deal with game states and Game Over conditions. These are all potentially more complicated concepts, but we'll start with simpler examples to master the concepts we'll need for our later projects. Conditionals, variables, and game states will allow you to go from simple button-press responses to complex and adaptive games as you grow familiar with the concepts.

DOI: 10.4324/9781003399018-9

Figure 9.1 The Butterfly Catcher project finished.

Motion. We'll work with more of the •Motion code blocks and a number of the hassle-free movement and position methods included in Scratch.

Mouse Reactions. The game will have the user click sprites with the mouse and have them react differently.

Difficulty Controls. We'll also include some control factors that will allow you to alter the difficulty of the game.

Text Display. We'll convey text and numerical data to our users in a few different ways in this game.

Lockdown. To stop things from reacting after the game ends, we'll use conditional statements to stop sprites from reacting.

Clones. To make the game challenging, we'll use clones to create more and more dangers to the player!

Sound Cues. We'll give the player sound effect cues when they score, lose a life, or the game ends, for more reactive engagement.

Variable Testing. Taking Ifs to the next level, we'll see how we can use • Operators and •Variables to create more dynamic conditional statements.

Player Lives. To make the game last a little longer, we'll see how we can give our players multiple lives so they can make some mistakes before it's game over.

Building It

Step 0: Create Your New Project

Make sure you're logged in to Scratch, then click Create to begin a new project! Since we won't be using it, we can delete the Scratch Cat sprite by clicking on the trash bin on that sprite's thumbnail in the Sprite Listing.

Step 1: Selecting the Background

Remembering what you've learned in the earlier projects, select a background for the project. One of the nature scenes is best. I used **Forest** in the examples.

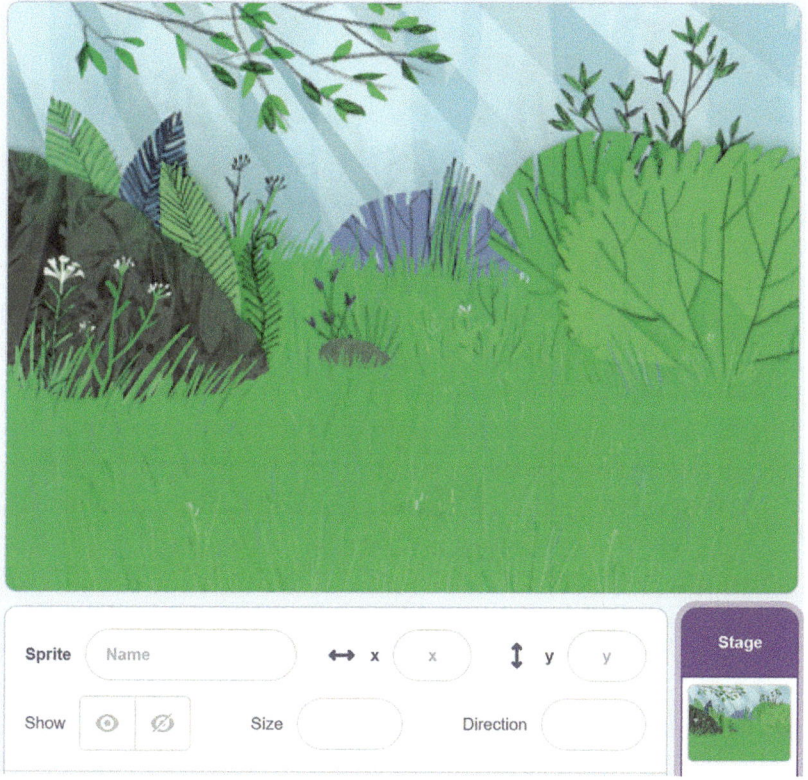

Step 2: Bouncing Butterfly

Try adding the *butterfly 2* **sprite** to the project. We're going to make it fly around the screen at random. If you don't remember how, check the earlier projects for instructions on how to add a sprite.

To get our butterfly moving, we'll, as always, start with an •**Event** block. Using the ❶ •**[When ▷ Clicked]** event, our butterfly will start moving as soon as the game starts. Next, we'll need a ❷ •**[Forever]** code block from •**Controls** to ensure that it will always keep moving.

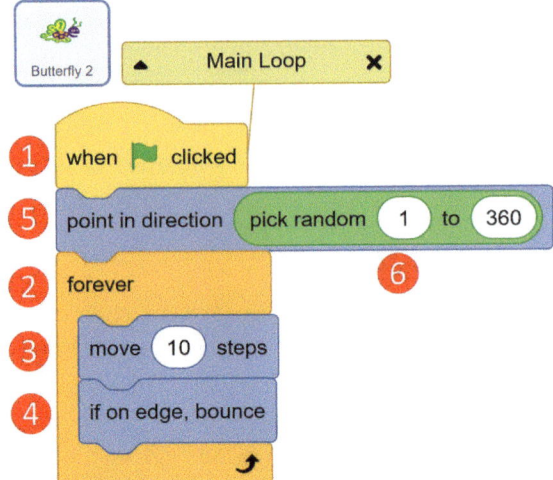

To get our butterfly moving, we'll need some ●**Motion** code blocks. Let's grab a ❸ ●**[Move (10) Steps]** and add it inside our ●**[Forever]** loop. If you test your game now, you'll get a butterfly that just flies straight to the right side of the screen but stops before going completely off-screen. Scratch tries to prevent things from moving off-screen – having them always visible and clickable so learners don't lose track of their sprites in game. If you drag the butterfly back to the left, as soon as you let go, it will fly right. That's exactly what our code is telling it to do – ●**[Forever {●Move}]**, with its direction by default facing right.

To get out of this tricky situation of them getting caught on the wall, let's add an interesting ●**Motion** block: ❹ ●**[If On Edge, Bounce]**. Place that in the ●**[Forever]** loop below the ●**[Move (10) Steps]**. Now you'll see your butterfly keep moving but bounce back and forth between the left and right walls. Progress, but still a little too easy and predictable, so we'll add one more complication.

Do you remember how we had our firework rotate to a random direction? Using that same technique, our butterfly can fly in any direction. Let's grab a ❺ ●**[Point In Direction (90)]** code block and add it above the ●**[Forever]** code block. By doing so, it will run when the game starts, but it won't run continuously the way anything inside the ●**[Forever]** code block does. You'll also notice that you can't have it below the ●**[Forever]**, since ●**[Forever]** never ends; nothing can come after it.

With our ●**[Point in Direction (90)]** in, we now need to randomize it. Go to the green ●**Operators** code blocks and grab a ❻ ●**(Pick Random (1) to (10))** and place it into the ●**[Point in Direction (90)]**. Now change the numbers from 1 to 360 and our butterfly will head in any direction. Test it out by clicking the ▷ to see our butterfly is now a lot harder to keep track of.

Step 3: Get Flapping!

Now that our butterfly is successfully flying around, let's get it looking like it's flapping around. Do you remember how we animated our dinosaurs in the first project?

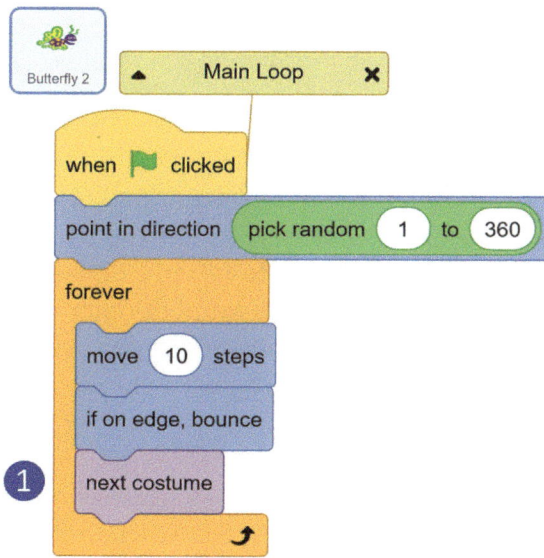

In the ●**Looks** blocks, grab a ❶ ●**[Next Costume]** code block, place that in the ●**Forever]** loop, and our butterfly will switch costumes with every step. Thankfully, the built-in costumes for butterfly 2 make it flap its wings. *Perfect!*

Step 4: Catching the Butterfly

We got our butterfly moving around; now let's make sure we can catch it. For this we're going to need a new ●**Event**. Grab the ❶ ●**[When This Sprite Clicked]** event for the butterfly. The player will catch the butterfly by clicking on it with their mouse.

So what happens when the player catches the butterfly by successfully clicking on it? Well, because we want our game to keep running, we aren't going to remove it from the game. Instead, we'll have it disappear and reappear somewhere else, so the player has a "new" butterfly to catch.

In the ●**Motion** code block, get a ❷ ●**[Go To (Random Position)]** code block. This is a very useful code block that handles randomizing an object's location on-screen in one easy code block. Try clicking on the butterfly now and you'll see it reappear somewhere else. Let's add one more complication here. The current problem is, with a random direction, it might still end up

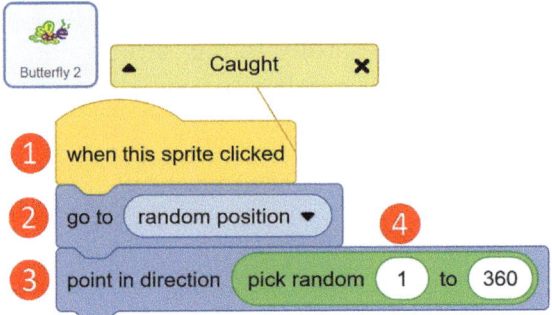

going straight up and down or left and right. What we really want is that when clicked, a new random direction is given, making it less predictable.

Just like in the •[When ▷ Clicked] code block, add a ❸ •(Point In Direction (#)) with a ❹ •(Pick Random (#) to (#)) placed below the •[When This Sprite Clicked], with the •Pick Random numbers changed from 1 to 360. You can grab new code blocks or try to duplicate the one you have already.

Step 5: Tracking Your Score

Our next task is to keep track of the player's score. For this we're going to need a •Variable. Go to the dark orange •Variables code blocks. Here we need to **make a variable** before we can use it. Click the **Make a Variable** button, and a pop-up will ask you to name it. We'll call this variable *"Score"*. We want the *score* visible, so we'll leave its checkbox in the Variables code blocks list checked. You can check the earlier Batty Flaps project for more information on creating a *"Score"* variable.

With the variable created, now we can add our code blocks. Add a ❶ •[Change [*variable*] by (1)] code block to the •[When This Sprite Clicked] event. Make sure it is pointing to the new *"Score"* variable. This block will take whatever the *"Score"* currently is and add 1 to it. Test the game. By successfully clicking on the butterfly, your score should increase by 1.

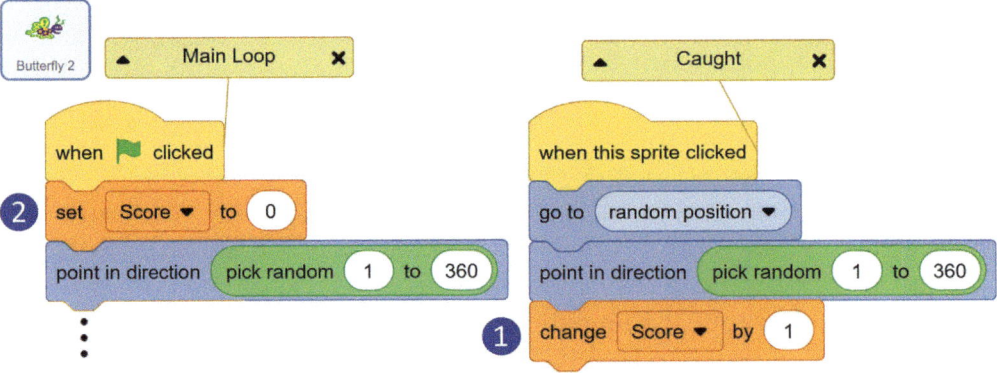

The one thing that might not be obvious is the need to reset your *"Score"* variable. Right now, if you stop the game and click the ▷ again, you'll start a new game, but the variable doesn't reset to 0. So let's add a code block to ensure that we'll always start the game with 0 points. Take a **2** •[Set [*variable*] **To (0)**] code block and add it under the •[When ▷ Clicked] event above the •(Point in Direction (#)). This way, the variable is set to the exact number 0 when the game starts. Keep in mind that we wouldn't want this in the •[Forever] loop, since that would constantly reset the score to 0!

In Book 3: Advanced, we'll take this to new heights by teaching you how to create your own worldwide high score leader boards to track scores of all your players on Scratch!

Step 6: Adding the Beetle

Now that we've got our butterfly acting right, let's make things challenging by adding in our first beetle. Add the *beetle* sprite from the library. Because we want multiple beetles, we're going to explore some basic cloning in this game. This means we're setting up our *beetle* a little differently to handle having multiple copies of it generate in our game.

In the beetle's code, let's start with a **1** •[When ▷ Clicked] event. Under this we'll have a •[Hide] code block from the •Looks category. For now, all the *beetle* does when the game starts is disappear.

Beetles are added to the game to get in the way of the player trying to click on *butterfly* 2. They'll have their own event if the player accidentally clicks them. For now we just need a basic reaction to signal that they got clicked.

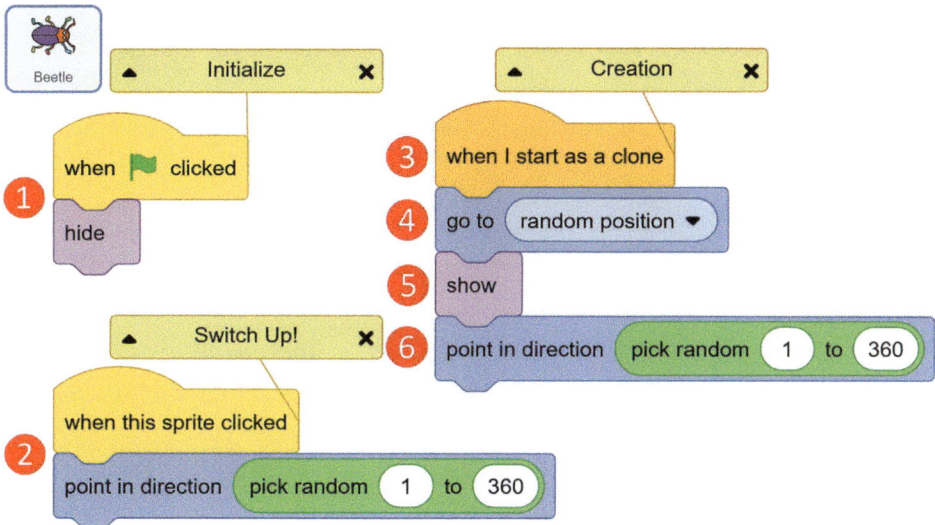

Add a **2** •**[When This Sprite Clicked]** event, and under it give it a •**(Point In Direction (#))** with a •**(Pick Random (1) to (360))**. For now, the *Beetle* will just change direction when clicked on.

The last set of code we need for now is handling creating beetles. As we mentioned, we're going to clone them, which creates a copy of the sprite in the game. We need an event, but the cloning-related events are in the •**Control** category of code blocks. Grab a **3** •**[When I Start As A Clone]** code block. This is the event that runs when a clone is first created, acting like a •**[When ▷ Clicked]** event for these late joiners. It's important to note that no clone will ever be affected by the •**[When ▷ Clicked]** event, since they only ever get created after that event has happened!

When a *beetle* is created, we want it to start at a random position. Add a **4** •**[Go To [Random Position]]** code block from the •**Motion** category. This will start them anywhere on the screen, but if you try creating a clone, you probably won't notice much because we used •**[Hide]** on the original *beetle*. All the clones will inherit that state, so let's add a **5** •**Looks** •**[Show]** code block to make sure our clones are visible. Lastly, we'll make sure the beetles start in a random direction; we can duplicate our existing **6** •**[Point In Direction** •**(Pick Random (1) to (360))]** combo and put the copy in this stack.

Now, how do we create some beetles? For doing so, we're going back to our *butterfly 2* code. In the •**Control** code blocks, find the **1** •**[Create Clone Of [*Sprite*]]** code block and add it to the bottom of the •**[When This Sprite Clicked]** stack. Make sure the type of clone created is a *beetle*. Now every time you click the butterfly, a new *beetle* should appear on the screen.

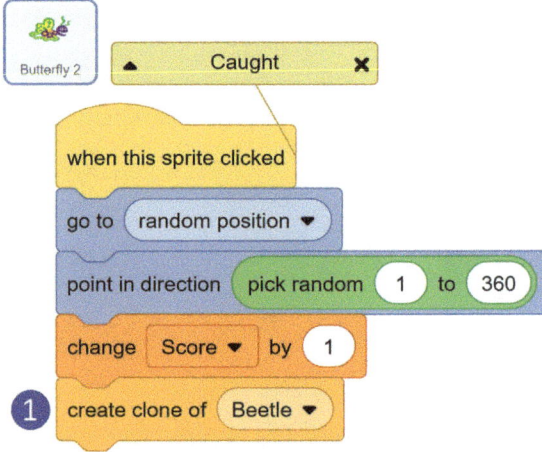

Step 7: Bouncing Beetles

With the *beetle* clones just sitting there, it isn't too hard to avoid clicking them. Let's raise the stakes and get them moving using the same technique applied on *butterfly 2*. In our *beetle's //Creation* stack, we'll need a **1** •**[Forever]** code block to keep them moving, and inside that a **2** •**[Move (10) Steps]**, followed by an **3** •**[If On Edge, Bounce]**. Our beetles should now all run around the screen at random. Any beetles currently in existence won't have run the •**[Forever]** loop when they first received the •**[When I Start As A Clone]** event, so it's important to do a stop-and-start whenever you make changes to make sure your projects are running the full current code.

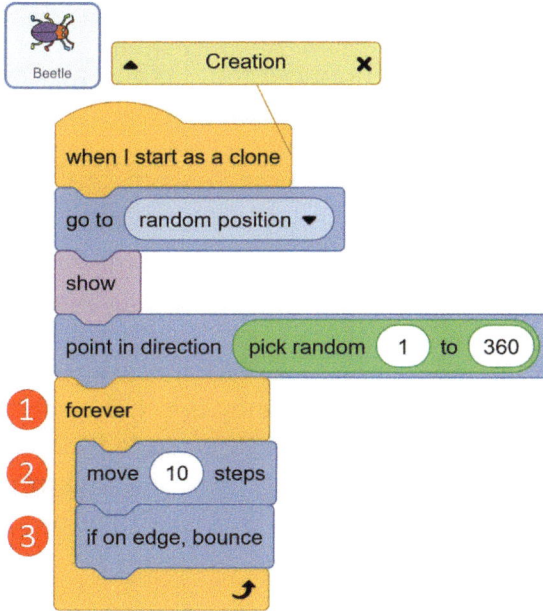

This is just an introduction to all the amazing things clones can do in Scratch. For more amazing techniques, check out the Snowball Fight game to make projectiles in Book 2: Intermediate, or the Scrolling Shooter project in Book 3: Advanced, to see how to make complex enemies for games.

Step 8: Player Lives

You can see how challenging the game is at this point, but there isn't really any consequence to failure. By adding in player lives, there'll be a risk of having an end to our game if too many mistakes are made (*beetles* are clicked).

Select the *butterfly 2* and go to the •Variables code blocks. We're going to need a new •Variable. Click **Make a Variable** and enter the name "*Lives*" for this new variable. To set player lives at the start of the game, add a **1** •**[Set**

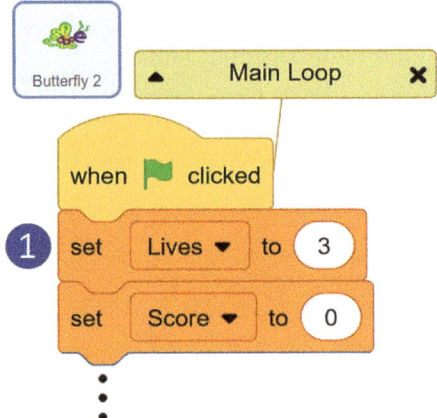

[variable] to (0)] code block right under the ●[When ▷ Clicked] and change the ●[*variable*] to *"Lives"* and the value to (3). The player will now start the game with (3) *"lives"*.

Now let's switch over to the *beetle*. Here we'll add a ❶ ●[Change [*variable*] by (1)] code block to the bottom of the ●[When This Sprite Clicked] stack. We'll change the ●[*variable*] to *"Lives"* and the value to (-1). This means the player will lose a life whenever they click on a *beetle*. Test it out.

You probably noticed the game doesn't stop when you have no more lives. To do so, we'll need to make a conditional statement. If the player has 0 *"lives"*, the game should end, but if they don't, the game should continue. A conditional statement needs at least three code blocks. First, in ●Control we'll get an ❷ ●[If <condition> Then] code block and put that at the bottom of our ●[When This Sprite Clicked] stack. This allows us to ask a question – if something is happening, if something is true. Our ●If question will be Boolean (true-or-false answer), allowing us to test something about the game. If the answer is true, it will run whatever code we put inside the ●If. If the answer is false, it will skip right over it.

Flow of Code

One of the major things we need to pay attention to when coding projects is how or when things run. If we attach code to a ●[When ▷ Clicked], it simply runs once and never again. If we put it in a Forever loop, it will run forever, but it will never stop running. Often, we want things to only run sometimes. Not once, not forever, but sometimes. This "in between" isn't obvious to achieve. There are no code blocks that tell something to only run sometimes. Instead, we need to build our own code to manage when sometimes is. The ●Control code blocks are largely about this. Code blocks like ●[Repeat (#)], ●[If <condition> Then], and ●[Repeat Until

<condition>] are ideal ways that we can have code run until a condition is met. As long as we learn to create Boolean (true or false) conditional statements, we can have something run only when that condition is true. We can then use our code to allow that condition to become true or false when we need it to. In this project, we'll have the game run while the player has lives left, and stop when they don't. In our other books in the series, we will explore other conditional statements showing how player turns can work, or how to implement different game modes/screens, among others. Conditionals are a vital skill to create dynamic and reactive projects.

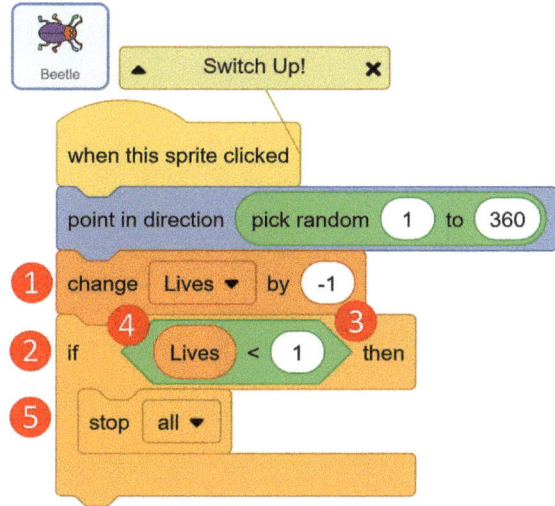

Let's figure out how to ask that question. Our next code block is in the ●**Operators** category. The ❸ ●**((0) < (0))** (lesser than) code block allows us to compare two numbers. But we can't just type in what we want. Any number we type in will always be the same, and we'll always get the same result. For example, 0 < 0 is always false, and 3 < 7 is always true. We want to test something dynamic, something that changes, so that our program will adapt to those changing conditions. If something is always the same, there'll be no reason for an ●**If** statement.

What we want here is to test our *"Lives"* variable. Going to the ●**Variables** category, we can see little oval *reporter* code blocks of any variable we've created. These are the placeholders for the value of the variable. We can take our ❹ ●**("Lives")** code block and put it in the first value of our Less Than comparison code block. Since we'll be testing if our ●**("Lives")** is less than 1, set the second value to (1). That way, our ●**If** will run its code only if we have zero *"lives"*!

Lastly, inside our •**If**, we need to place a •**Control** code block ⑤ •**[Stop All]**. This will essentially end the game by stopping all the currently running as soon as the (*"Lives"* less than 1) condition is met.

Step 9: Beetle Game States

At this point, by testing the game, you probably noticed two things: (1) If statements are great, and 2) our Game Over is a little abrupt and uninteresting. Let's use our latest findings and employ •**If** statements to set up a better ending to our game. We'll use a conditional statement (an •**If**) to change how the game runs, handle a Game Over state, and make our beetles pause while displaying information to the player in place of the current abrupt game end.

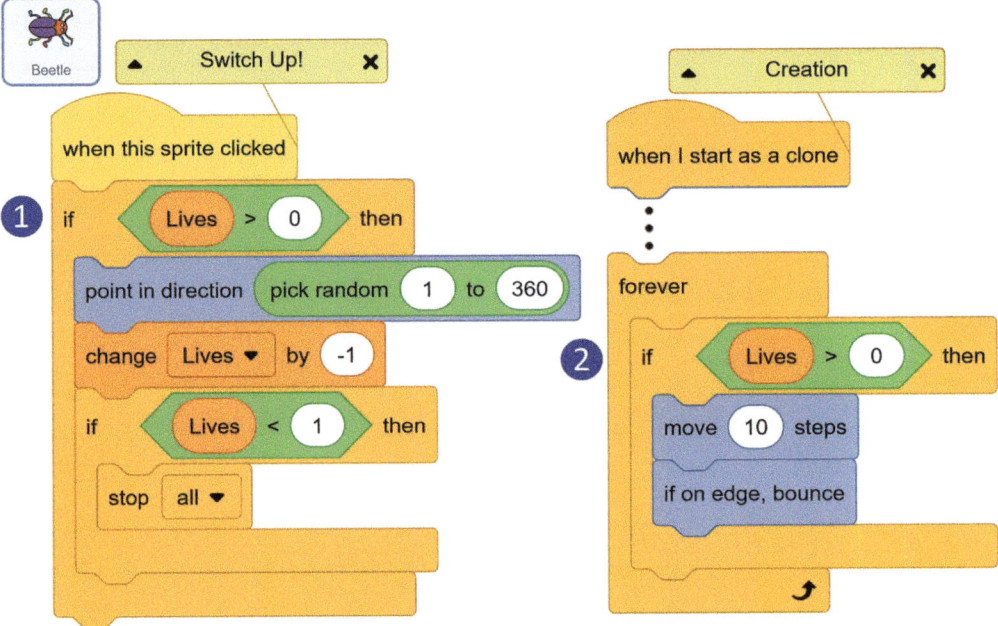

We're going to need two copies of the following •**If** statement. I suggest creating it first in the empty white space in the *beetle* (unattached to any existing code block at first) so you can duplicate it before placing into the appropriate locations. Start by getting an ① •**[If <conditional> Then]** from •**Control**. Next, we'll add the •**((0) > (0))** (greater than) code block for •**Operators** inside it. Lastly, add the •**("Lives")** variable into the first number and type "0" into the second number. Our •**If** will run only when the player still has *"lives"* left. Make a duplicate of this ② •**If** statement. Use them to limit the *beetle's* code so that they only move while the player has *"lives"* and only respond to being clicked when the player has *"lives"*. By nesting the normal functions into these conditionals, we are making something called a "*game state*". Our program is detecting a given condition and changing how it functions in response. You won't see much of a difference in the project just yet. Don't worry.

> *Game states, or state machines, like this are an amazingly important and powerful technique to know. We'll take this concept a lot further in projects like our Big Map Racing game in Book 2: Intermediate or the Point-and-Click Adventure in Book 3: Advanced.*

Step 10: Game Over Announcement

Now that we've got the *beetles* stopping when the player loses their last life, we can add in some code to make a better Game Over moment. Head to the ●**Looks** code blocks to add in some information displays.

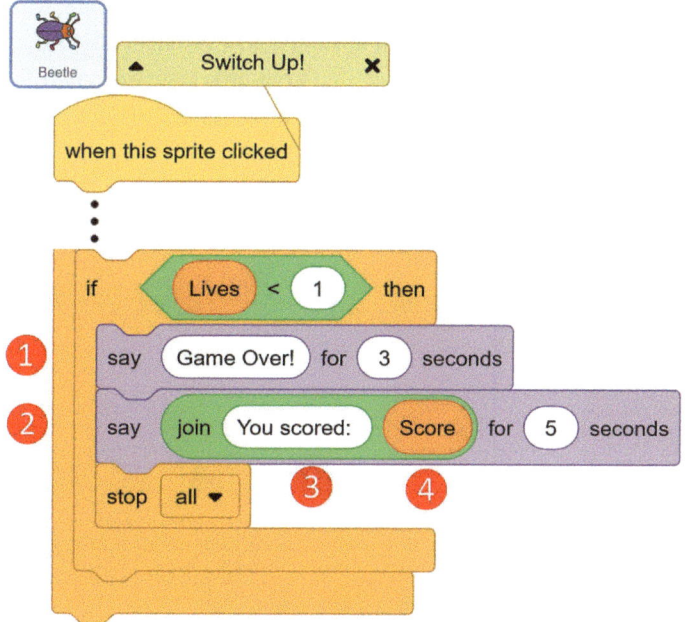

Grab a ❶ ●**[Say (text) For (#) Seconds]** code block and place it above the ●**[Stop All]** (included on the ●**[When This Sprite Clicked]** script). Change the text to "Game Over!" Next, we'll add the final score. To do so, start by adding another ❷ ●**[Say (text) For (#) Seconds]** code block below the first. Here we're going to make a slightly more complex code to display the message. Start by going to the ●**Operators** code blocks and getting a ❸ ●**(Join (text1)(text2))** code block. Place it in the text for the second ●**[Say (text) For (#)**

Seconds] code block. The ●**(Join (text1)(text2))** code block allows you to join together two different parts into a single message. Why would we need this? This way, we can incorporate our ●**("Score")** variable as part of the message. Add the ④ ●**("Score")** code block to the second part of the ●**(Join (text1) (text2))** code block. In the first value, type in "You scored: " (with a space at the end). The ●**(Join (text1)(text2))** block will take the number of the "*score*" and append it to this message. You may want to adjust the durations of the ●**Say** blocks for readability.

Try it out! You should see the *beetle* tell you your final score when the game ends.

Step 11: Butterfly Game States

While our Game Over is definitely better, our *butterfly 2* hasn't gotten the message. So let's add in our game state conditionals like we did for our *beetle* to the *butterfly 2*. We'll need our ① ●**[If ●<●(Lives) > (0)> Then]** conditional to control the code inside the ●**[Forever]** code block, and another to control all the code within the ② ●**[When This Sprite Clicked]** event. See if you can remember how from our *beetle*. You can review step 9 if you forgot how, and just apply it to the *butterfly 2* instead. You'll need both the inside of

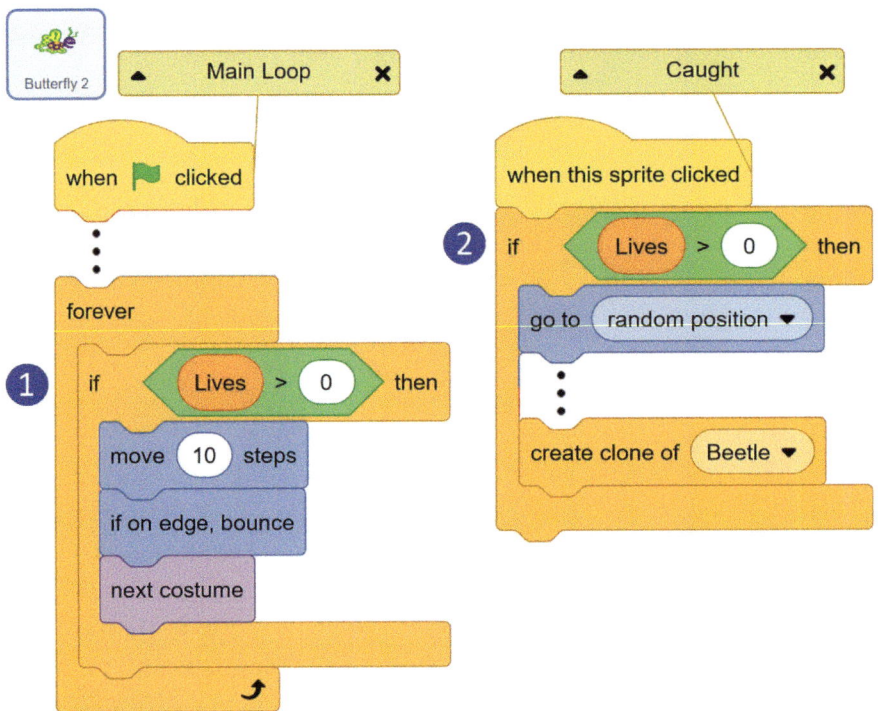

the •[**Forever**] block and the code in •[**When This Sprite Clicked**] contained inside conditionals to properly stop the butterfly when the game is ending.

Traffic Control

I just ordered it to •[Stop [All]], but I can still move the character? It can be a bit counter-intuitive when this happens, but it is intentional. The Stop code blocks don't just stop the game like we expect them to. They do exactly what they say, but not what we think. When you order a •[Stop[All]], all current running code is stopped; this means any •[Repeat (#)], •[Forever], or other loop stops functioning. The other options will •[Stop [This Script]] or •[Stop [Other Scripts In Sprite]]. But remember that Scratch itself is still running. This means when an event occurs, the code attached under any matching •Event code block will be run. Stop blocks stop code that is currently running, but it doesn't stop it from running again. In our earlier Batty Flaps game, we used this re-running feature to make our game replayable. With this game, we've made a hard lock so you could see two different ways to handle this issue, so you'll have both concepts to work with when you're making your own projects.

Step 12: Assessment

As a last tweak to our Game Over system, we can evaluate how the player did. In our *beetle*, we'll use •**If** statements to handle this task just above the • [**Stop All**] code block. Perhaps the player didn't do very well, and we want to contextualize their score. Get ❶ an •**If** statement from •**Control**. Inside, place (from the •**Operators** category) a •((0) < (0)) (less than) code block. Now we'll test the player's "*score*" by getting a •("*Score*") code block from • **Variables** and place it in the first value. Type in the second value as 10. This will run if the player scored 9 or less. From •**Looks**, get a •[**Say (text) For (#) Seconds**] block right under it. We'll add the text "*Better Luck Next Time!*"

Next, let's point out if the player did well. Get ❷ an •**If** statement and add a •((0) > (0)) (greater than) code block. Again we'll add the •("*Score*") **variable** to the first value. Set the second value as 9. If the player scored 10 or better, they'll get this message. Add in a •[**Say (text) For (#) Seconds**] block and change the message to "*Wow! Good Job!*" This should give the player some feedback on expectations. Of course, we could change these numbers for a different expectation, and if needed, the duration of the **Say** blocks can be increased for younger readers.

Step 13: Sound Effects

Lastly, we can add a few sound effects to our game, making it more dynamic and appealing.

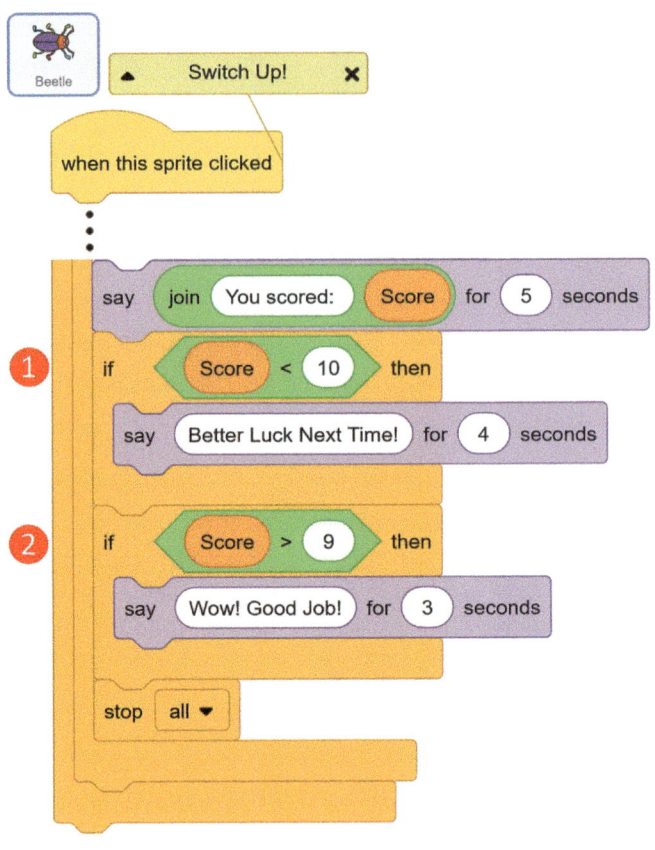

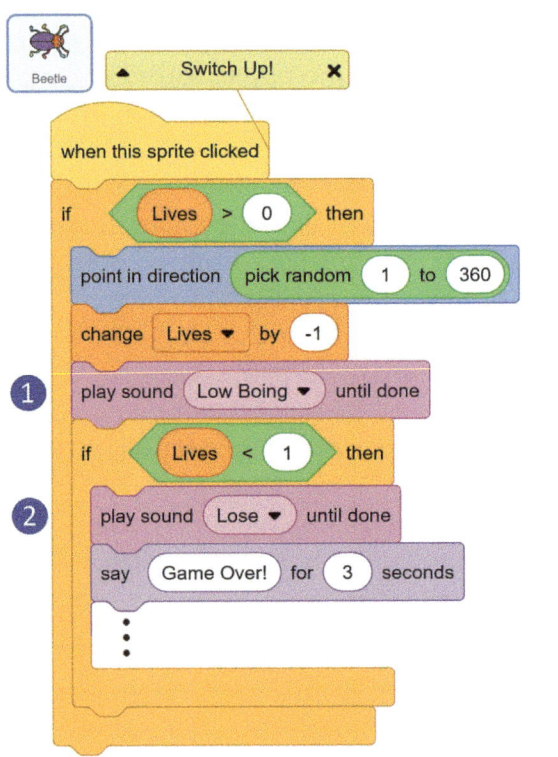

In the *beetle*, let's grab two **sounds** in the **Sound** Tab. We'll add the **Low Boing** and the **Lose** sound clips from the **Sound Library**. Switching back to the **Code** tab, add the ❶ •[**Play Sound (sound) Until Done**] code block below the •**Change Live**s and set its sound clip to **Low Boing**. This will give us a negative-sounding feedback SFX when the player clicks a *beetle* and loses a life.

Take a second ❷ •[**Play Sound (*sound*) Until Done**] code block and add it to the •[**If** •<•("*Lives*")<(1)> **Then**] block. Switch its sound clip to **Lose**. This will play a sad little sample to indicate the game is over.

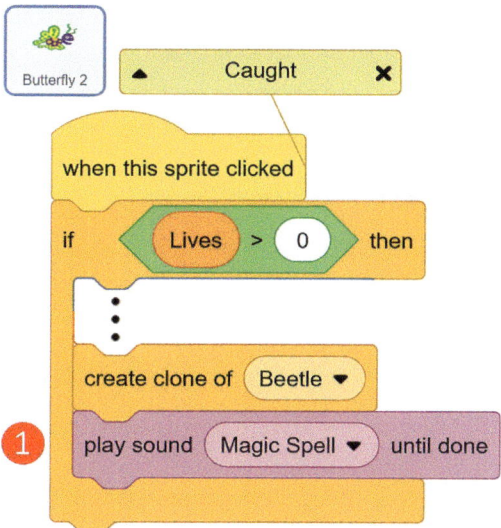

Lastly, we'll add a sound effect to the *butterfly 2*. Switch to the *butterfly 2* and go to the **Sounds** tab. Add the **Magic Spell** sound clip. In the **Code** tab, we'll add a ❶ •[**Play Sound (sound) Until Done**] code block inside the •[**If** •<•("**Lives**") > (0)> **Then**] conditional in the •[**When This Sprite Clicked**] event. Select the **Magic Spell** sound clip; this will play a happy little notice when the player successfully catches the butterfly.

Great work! Our game is complete! Play it to see all the moving parts at work. By the way, what's your high score?

For working copies of this and every project in the book series, visit www.massivelearning.net *for direct links to Scratch Projects, and to see our other projects and resources for coding education!*

10

Beginner Check-In

Now that we've completed the four beginner projects, let's stop and reflect on our progress.

Key Skills

By working our way through the four beginner projects, we've gotten a lot of practice with Scratch. Any time spent with Scratch is valuable at this point. Moving from beginner to intermediate is really about spending time and getting familiar with Scratch. Once you're familiar with where things are and what they are, we can move on to how to use them and strategies for hypothesizing and synthesizing your own ideas. These four projects were chosen to give you a wide range of experiences by letting you get familiar with the interface, the methods, and a lot of the most used code blocks and coding concepts.

Let's think about some of the core skills you've learned and practised so far.

Navigation and Interface

Our first task working with Scratch is simply understanding the interface. Versatility and utility often come at a cost of complexity. We needed to understand all the different sections of the Scratch editor to be able to use it to its fullest. Thankfully, our projects covered a wide range of possibilities.

DOI: 10.4324/9781003399018-10

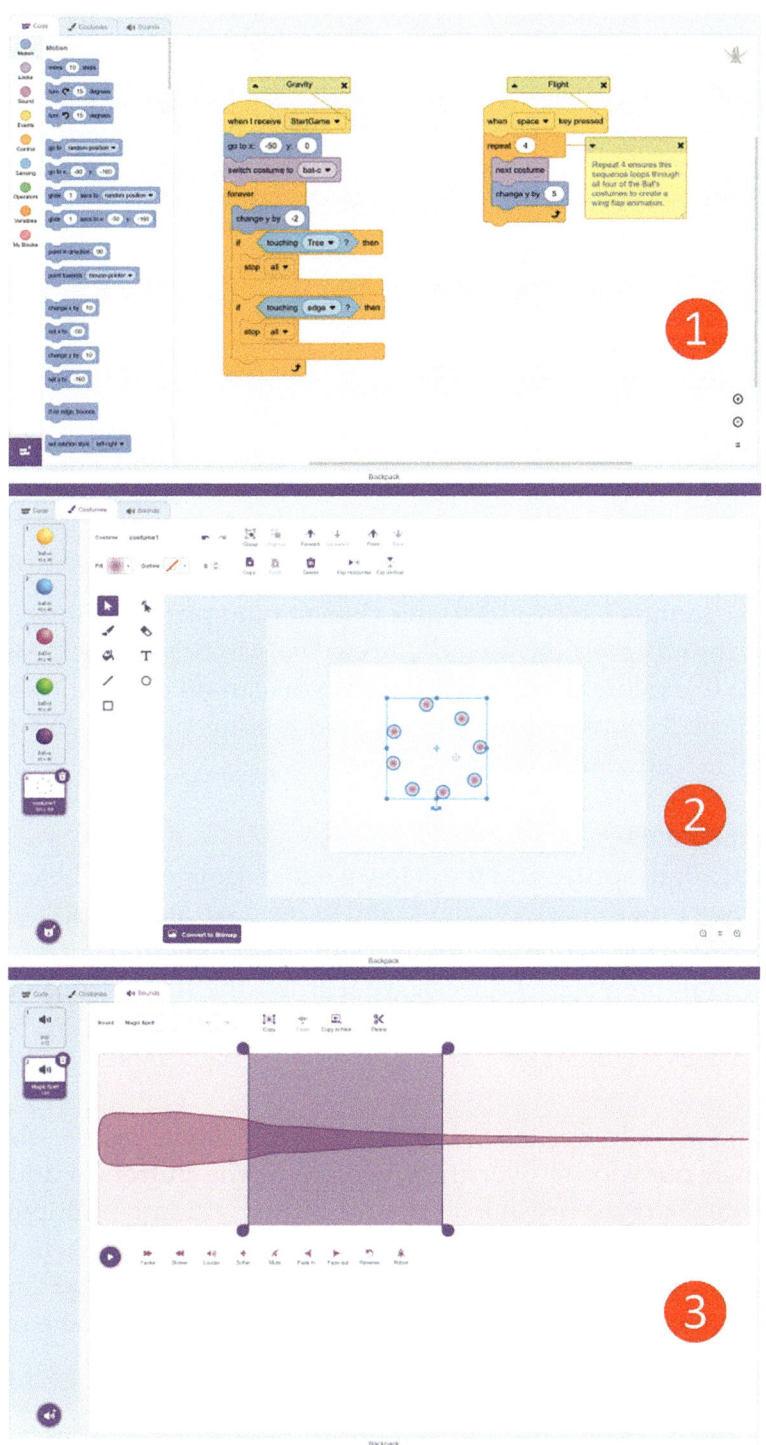

Figure 10.1 The three tabs ➊ Code, ➋ Costumes, and ➌ Sounds.

Code Tab

Most of our work was done in the Code tab. We learned how to use code blocks, how to delete them, how to connect them together, and how to break them apart. We learned the different categories of code blocks and how to navigate between them. We also learned convenient techniques for duplicating code and copying it to other objects or the Backpack.

Costume Tab

Working in the Costumes tab, we learned how sprites can have multiple costumes, allowing us to change how they look for animations or displaying different states. We learned how to add costumes and how to create our own costumes or alter existing ones. We got some practice with various art tools and art tool properties, the canvas, and how sprites display and position in Scratch.

Sounds Tab

We used the Sounds tab to add sound clips to our sprites for music or sound effects. We didn't cover the sound editing tools in Scratch in these projects, but feel free to explore them and see what you can do on your own. You can check out *Book 3: Advanced* for some sound editing practice and tips in the Point-and-Click adventure game.

Objects and Backdrop

Working with both sprites and the stage, we manipulate these objects to make our interactive experiences, displays, and games. We learned some of the differences between the two, tried out the built-in options, and custom-built our own. With multiple-object projects, we learned the importance of selecting the right object to apply code to.

Adding Sprites

Not only did we add sprites using the built-in library, but we also learned how to hover our mouse over the Choose a Sprite button to add our own custom sprite using the brush icon "Paint" button. Adding multiple sprites to projects, we learned to differentiate art and code assigned to each. We moved sprites both as designers with the mouse and through code and learned how to allow users to move objects with inputs or drag them with the mouse.

Adding Backdrops

Dealing with the Stage, we added built-in backdrops. Through the Backdrops tab, we made our own, albeit simple, backdrop. We also took a built-in

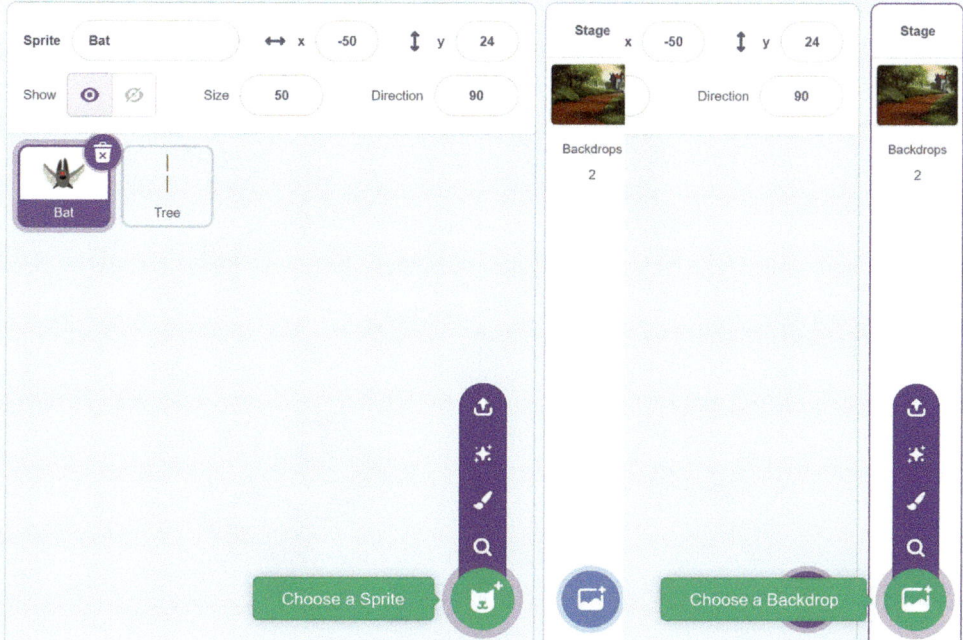

Figure 10.2 The Choose a Sprite and Choose a Backdrop buttons with their hover menus extended. This provides alternate options of uploading art, selecting a random piece of art from the Scratch Libraries, painting your own art, or searching for a specific piece in the Scratch Libraries.

backdrop and customized it using the art tools while learning to use the powerful Reshape tool.

Drawing Your Own

By choosing the Brush option when adding a sprite, we learned to draw our own custom sprites and costumes. We adapted built-in sprites by adding additional costumes to them. Taking a premade background, we learned to add to it and customize to our purposes.

Object Properties

By working with sprites, we learned the concept of object properties – the data points that tell Scratch how to position, move, and display the sprite. We worked with these both directly through editing them in the Properties panel, using code blocks to directly change them, or indirectly by using code blocks to achieve a purpose (like Move Steps) and have them change as a consequence (X and Y coordinates in the case of Move Steps). Object properties are a very important coding concept that will only get more relevant as you continue to code.

Figure 10.3 A close-up of the Stage Window and Properties panel. Can you figure out which Scratch Cat is selected and where it is in the Stage Window?

Coordinate System

In Scratch, all objects have X- and Y-coordinates. We learned to work with these both directly through coordinate-specific code blocks and consequentially through simply moving objects around the Stage Window with the mouse, or through movement commands. Scratch can be used to explore four-quadrant coordinate planes with students. But we don't have to, since the computer will track coordinates even if we're just using more human, or centric, movement commands like Turn Right and Move 10 Steps.

Sizing

We learned about object's sizes and how to draw them in-game large or small, regardless of their costume. We directly altered this value in the Properties panel, as well as dynamically in code. We also learned how coordinate positions and movement to coordinates are unaffected by size. Unrelated to objects, we also explored how size affects the brush and outline in the Costumes tab.

Direction

Using Turn blocks, we explored sprite's facing, which allowed us to explore geometry and drawing. We didn't use direction too much except with Turn blocks so far, but we'll explore some other interesting new options in future projects.

Visibility

We worked with the ●[Show] and ●[Hide] blocks to make objects appear and disappear. We also learned that how code runs can be affected by the *visibility* property, and used the ghost effect to play dynamically with visibility by changing an object's transparency.

Costume Changes

Working with the Costumes tab demonstrated how sprites can have multiple costumes assigned to them and that they have a specific costume they get drawn as at any given time (when visible). We set costumes to specific ones or simple cycled through them with the ●[Next Costume] code block. We added and edited costumes as well as used the built-in ones from the library.

Graphical Effects

Exploring the Graphic Effects functions in Scratch expanded our options on how to draw sprites. We used the ghost effect to change the transparency and learned the importance of ●[Clear Graphic Effects] to undo changes.

Sequencing

Coding is how we give computers instructions. It needs to be clear and concise. While Scratch gives you a lot of flexibility in the workspace, code is still executed in an orderly fashion. Learning how code works as a sequence of instructions is a useful thought process that can help understand some of the more complex coding techniques and be a handy tool for understanding and correcting bugs.

Sequences

At the most basic level, we learned to make sequences of code. A sequence of code blocks is made by snapping them together so it can be read from top to

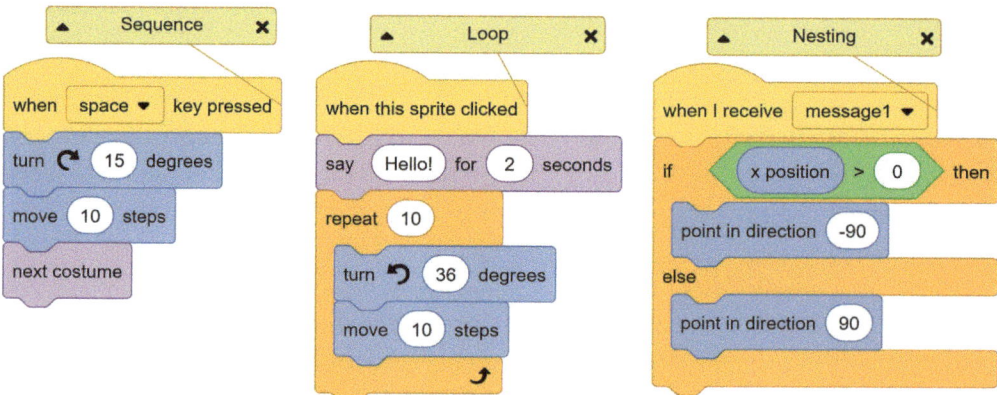

Figure 10.4 Here we see an example of a basic sequence of code, an example of a loop using the Repeat block, and an example of nesting using and If/Else block, which is also a conditional statement. These four concepts are critical components of any computer science.

bottom (by humans and the computer). Through practice we learn how code blocks can affect each other. Sometimes the sequence needs to be absolute, as in one block must follow another. In other cases, they can be in a varying order without affecting the results. By understanding sequences, we can move beyond just understanding when code will trigger based on their event, and comprehend the process, order of operation, loops, calling, concurrency, and other important considerations. But don't worry about all that now. We are gradually establishing the foundation of understanding how computers think and work.

Loops/Repeating

Using the •[**Repeat (#)**] and •[**Forever**] code blocks, we set up loops, ways code can run multiple times, without requiring an event to trigger each time. Loops are a powerful tool to ensure the functionality of games or projects (such as our use of the •[**Forever**] loop to keep our Dinosaur Dance Party rockin'), as well as a handy measure to do something a sufficient number of times (such as repeating our fireworks movement up into the night sky). Loops keep Scratch running a specific section of code, frame after frame within a stack, instead of running the whole stack each frame. Sometimes this is a benefit, sometimes a hindrance. By understanding loops, we can expect the correct result. Sometimes, code working over frames can be used to our advantage (to make an animation), and sometimes it makes for unfortunate delays (we avoided this in our examples, but imagine asking a computer to process a large database and only having it manage a single calculation per frame instead of the millions of calculations a second they are capable of). We'll learn a lot more about working with loops in the other books in the series.

Nesting

Through the •[Repeat (#)] and •[Forever] loops, as well as the •[If] block, we learned about the concept of nesting or nested events. The C-shaped code blocks allow us to place other code blocks inside them (nesting). Here, the outer code block controls how the inner code blocks are run, and this is why most of those kinds of code blocks are in •Control. You can think of these code blocks sort of like an order of operation parenthesis, meaning, they tell the computer to resolve whatever is inside before proceeding. Sometimes you nest code blocks inside other nesting code blocks, creating complex, almost-clockwork-style machinery (you'll see some great examples of this in our Pen Tool Fun project in *Book 2: Intermediate*). Nesting controls can create game states like in our Butterfly Catcher game, having the game function differently when the player had lost the game. In our more advanced projects, we'll use lots of examples of nested events to control when or how code runs.

Triggers

All our code is assigned under events. •Event code blocks determine when code triggers, meaning, the input or circumstance that calls them. We only used a few of the events: •[When ▷ Clicked], •[When This Sprite Clicked], and •[When (@) Key Pressed]. These cover the most useful and common triggers, but we'll learn a few other useful tricks to call code in future projects.

Animation and Timing

Animations let educators and students bring things to life – stories, characters, even scientific principles. We explored some of the key concepts and methods with our four beginner projects while exercising core skills that enable the creation of a range of possible projects.

Frames per Second (FPS)

One of the most important concepts one can grasp when working with Scratch is the frame rate. The term from film can be a good way to describe it since we (at least most of us) can think back to the actual physical film reels to understand the concept. As Scratch (or any programming language or editor) runs, it executes the code commands, but it has to relay the information to the monitor to convey visual data. This creates unique single drawings representing the state of the project. Each sprite is drawn in its current position, size, etc. on top of the current background. This snapshot, or frame, is then displayed on the monitor. The game then proceeds and updates whatever is happening in the project, moving sprites, changing their appearance or whatever the code entails, and then creating a new drawing to display. Each drawing displays for 1/30th of a second, then changes to the next one. Just like a

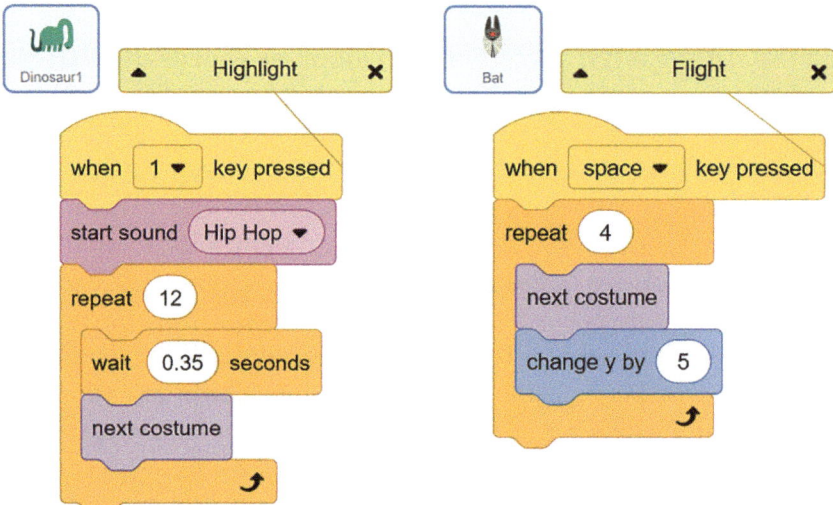

Figure 10.5 Two examples of animation code, the first from our first project using a delay for costume changes, the other from the third project changes costumes while also moving the sprite.

movie screen or many TVs, 30 times a second our game updates, making our brain interpret these rapidly changing drawings as a moving picture. Importantly, Scratch is running all the graphical code at 30 frames per second. This ensures everything is tied together seamlessly in coordinated, synchronized sequence. If no graphical changes are being made, the code can run faster than 30 FPS, so don't count on it, unless you're certain!

With some code blocks we break out of the standard 30th of a second frame rate. Code blocks that mention time intervals will execute for that time period without Scratch moving on or ahead. When a ●**[Wait 1 Second]** code block comes up, Scratch won't run the code after it until a second has elapsed. With code blocks like this, we can adjust the flow of time in our project.

With nested code blocks like repeat, each repetition runs in a separate frame. We can use this delay to ensure animations look great, with each step in time executing one iteration of the animation. As long as we expect this behaviour, we can use it to our advantage. Later on, we'll learn a secret technique to get around this delay whenever we need to.

Motion

With an entire category of code blocks dedicated to it, ●Motion is very important in Scratch. We've learned to use different forms of movement to ensure our sprites are where we want them, facing the right direction, and moving in the right way. Through code-based movement, we can simplify animation by using repositioning, scaling, and rotation handled in code, rather than having to make unique frames of animation for every change. Traditional animation doesn't have that luxury.

Looks

The key to most animation are the ●Looks code blocks and the costumes of a sprite. We learned to add or alter costumes, but also a number of important ●Looks code blocks that can be used to change the appearance of things. Changing costumes can unlock all the possibilities of the art editor, but through the graphic effects in Scratch, useful tweaks and reimaginings of the same graphical data can also be added.

User Input

Interactivity is what makes Scratch such a powerful tool. Instead of making set and permanent displays, documents, or other media like movies, Scratch can allow users to interact with creations. While we don't have to use interactivity, and we can simply make set animations, movies, or literature, the User Input functions allow us to work with our users. The inputs themselves are relatively simple, but what they enable can be a wonderfully complex and capable medium.

Key Presses

Our most used method of input were key presses. We learned how we can have the user enter commands via their keyboard to make things happen.

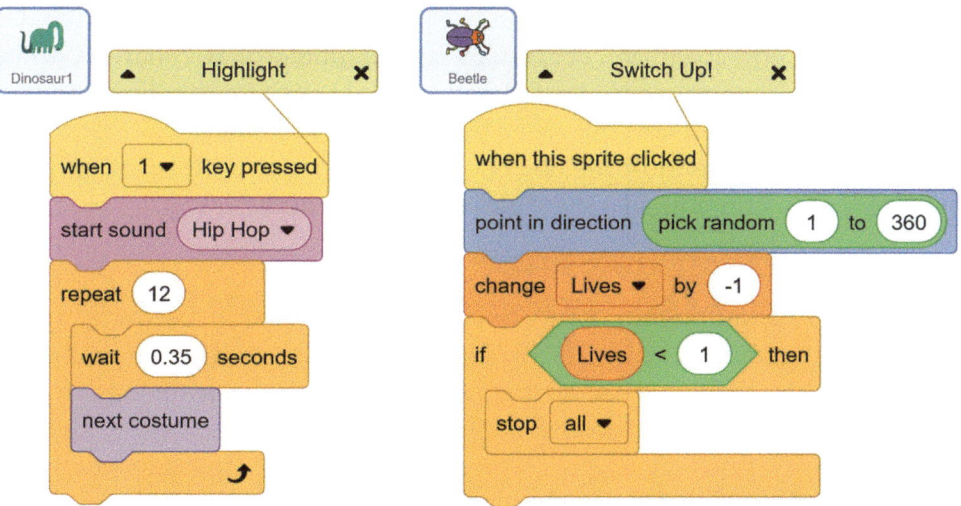

Figure 10.6 Two examples of user input: The first, from project 1, activates on the user pressing a keyboard button. The second, from the fourth project, runs when the user clicks on the sprite.

While simple to execute, the ●[**When (@) Key Pressed**] opens up a lot of input options. We'll learn in later projects some more complexities to working with key presses.

Mouse Input

We didn't work too much with the mouse yet, but we did allow for a few uses. We learned to have click-reacting objects so we can have clickable reactive components in our projects. We'll give some hints about how you can use the mouse for even more input and control for users in the later books in this series.

More Beginner Practice

If you aren't feeling confident yet in the preceding principles, you should try creating more beginner-level projects. Even if you are confident, practice always helps! Here are three suggestions for other projects you could try to make.

Apple Catcher

Apple Catcher is a classic first project for Scratch. You need a bowl that the player controls and can move left and right. Apples fall from the top of the screen. Each one you catch earns you a point. The game ends when you miss an apple and it hits the bottom. Can you figure out how to make it on your own?

Once you've made an Apple Catcher game, try mixing it up a little. Can you figure out how to control the bowl with the mouse or keyboard? Can you have the game continue if apples fall, but instead introduce an item you don't want to touch that will end the game instead (for example, a wasp or a rotten apple)? Can you have multiple object types that give you different points for catching them? Can you add in a "player lives" system?

Clicker Game

A surprisingly popular game genre on Scratch is called a Clicker game, named from the first game in the genre, "Cookie Clicker", as well as the main interaction element – clicking on something. A clicker game basically has an object you want to click that gives you some kind of resource or currency. Then you spend it to buy skins (cosmetic changes to the game), upgrades, or opportunities that let you gain more resources. There are lots of little variations and additions you can add into your game to grow it. You'll need a clickable object and a currency to start. But take a look at other examples and see what else you could add in – auto-clickers that generate currency on a timed event, multipliers, rare bonus clickable objects, a store to buy your upgrades, purchasable skins, etc. It's such a basic premise that you'll be able to theme it in just about any direction you want.

Birthday Card

Aside from games, you can also make some more artistic/presentation-focused projects. Try making some interactive birthday or greeting cards. Can you show an outside and inside of the card? Can you add in some music? What about some animations or effects? This should help you practice your art and animation skills in Scratch.

Teaching Beginner Scratch

You want to be confident and capable with Scratch in your own right, but the real goal is teaching it and using it in the classroom. Now that you've done the beginner projects, reviewed the key outcomes, and given yourself some extra practice, let's think about how we'll use this in the classroom. We want to make sure we're getting beyond just the rote copy-what-I-do follow-alongs; ideally, we'll be able to explain the theory and methods, as well as fix bugs and answer questions that come up.

Navigation

Your first job is to be a navigator, helping students explore the space and find what they need. Can you guide students around the interface? Do you feel comfortable with where to find things? Do you know what the icons are and where to go for what you need? If we can't clearly and effectively communicate where they need to go, students can quickly become lost, with a wrong click making it even harder for them to find their way back and then to where they wanted to go. Practice describing where you're clicking and where you're going as you work in Scratch. Once your students are familiar with Scratch, it won't be an issue, but you need to get them that comfort and familiarity with the interface before they can easily follow you through a project, let alone really start creating their own projects.

Think about how you add backgrounds and sprites. How to describe the code block categories and shapes. How they can handle code blocks, getting them, disposing of them, joining and splitting them, duplicating or copying them. How and when to switch tabs or sprites and switch back. Good, clear instructions on these tasks will help familiarize your students so you can focus on the more interesting and meaningful aspects, letting you teach *with* Scratch instead of just teaching Scratch.

Selections and View

Ensuring our students have the right view in Scratch is a big part of the battle. If they can't see what they need, they can't do what we ask of them. Most of the early frustrations with Scratch are a matter of view and selection. Clicking on the wrong thing makes it easy to get lost and not know how to get back. We can remind ourselves to highlight a few important things to help them find what they need.

Finding code blocks is a pretty big part of working in Scratch. Once you're proficient, you hardly notice it, but starting out can be a bit daunting. It can be very useful to highlight the code block categories (●Motion, ●Looks, ●Sound, ●Events, etc.) and the colour coordination system. Students should be shown they can use the category shortcuts on the left-hand side, but also that they can scroll the code blocks freely. You can talk about the colour coordination of the code blocks to help them find code blocks they see being used, and you can talk about the shapes, helping them understand how the code blocks relate to each other and fit together.

One of the biggest lessons for young students is understanding the object model in Scratch. Distinguishing between the backdrop and the sprites is very important, as many code blocks aren't available in the backdrop. If they complain they can't find a ●Motion block, that's a big clue that they've probably got the backdrop selected instead of the sprite you wanted. After adding a second sprite to a project, it can be good to stop and practice clicking on the different objects in Scratch to reinforce that lesson, so you can avoid a lot of trouble.

Other issues with selections and view are the level of zoom in their code view or their canvas view. The work area can be zoomed in, zoomed out, or returned to 1× zoom via the +, −, and = buttons in the bottom right. You can again focus on this to ensure students understand these functions, along with the work area scroll bars to move around their view so they can zoom in or out as they need and still see all elements being worked on. Sometimes students have really unusual bugs, and a lot of the time I've found the problem by zooming out and finding code block stacks they forgot they had placed. You'll also probably need to zoom in on your own code when you're screen-sharing with class, to make sure students can read your code blocks.

Code Order and Flow

Instructions need to be in order to get the right result. Understanding and recognizing the order and flow of code is important for building proper algorithms for the computer. We need our students to understand when and why sequence is important. A good example is the difference between a ●[**Move (100) Steps**] followed by a ●[**Turn Right (90) Degrees**], and a ●[**Turn Right (90) Degrees**] followed by a ●[**Move (100) Steps**]. The difference between the two is very important. This can be a great way to get students to appreciate the importance of sequence. By being able to explore and make mistakes (or just experiment), the value of choices can become apparent.

You want to be able to explain sequence to your students, and the preceding example is a great foundation for that. However, you also need to cover other complexities in the flow of code, where sequences get a little less simple to follow. We can start by ensuring students understand the ●Event triggers. The ones we've covered in our four beginner projects are fairly obvious, but future ones will be a little more challenging. Being mindful of event triggers can be a useful skill in understanding bugs and solving problems, so it's worth talking through these things. We might feel like we're stating the obvious with simple code, but we're priming a conversation that will deepen as they explore new concepts and complications. ●[Repeat]s and ●[If]s can be useful to have students show they understand the flow chart sequences that are occurring in their code. If our students can work through their code without it running by verbalizing the states and conditions and showing us the order and effects even with the simplest code, we begin a process of having them fundamentally grasp the logic, and they'll be able to expand that fundamental knowledge easier. It is important to do so before attempting it with more complicated models later on.

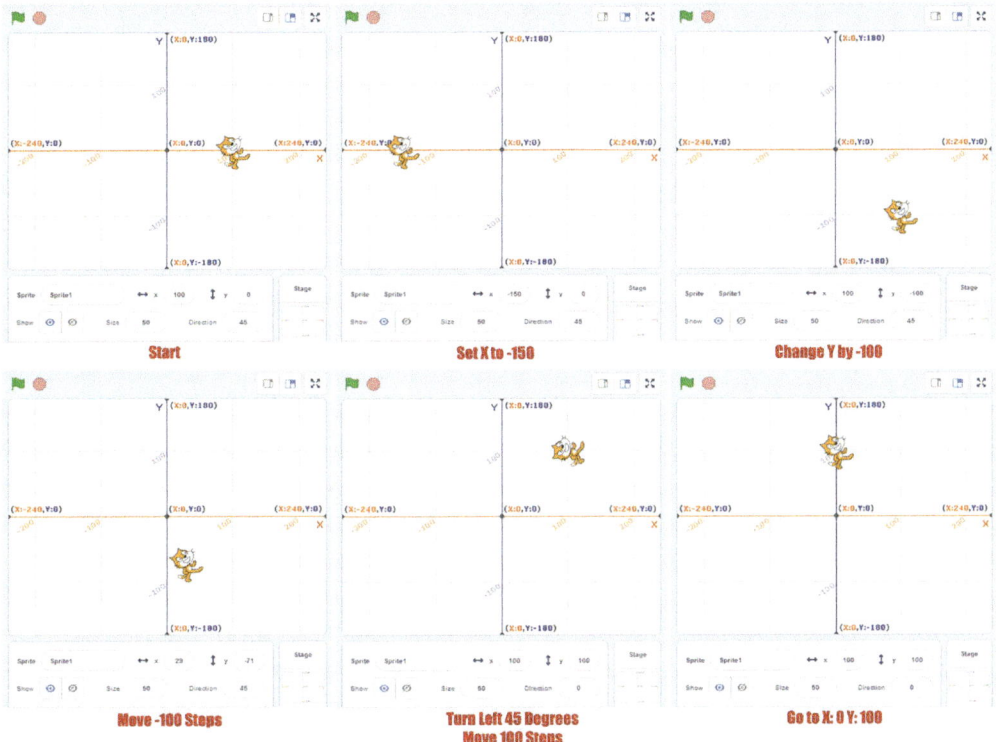

Figure 10.7 An example of movement in Scratch. From the Start position, each image shows the changes that will occur if given the commands listed below it.

11

Follow-Up: Extending the Projects

You may have noticed while building the projects that some things could be done differently. Maybe a feature you thought should be there was missing, maybe the myriad of techniques we showed could be simplified to your one favourite method, or some other changes. These projects were built with plenty of room for improvement in mind. Some were simplified to fit the learning curve, or to provide a diversity of concepts, or even to fit a more linear path of development while retaining usability at every stage. There were a lot of considerations to make while building the example projects, and having them calling out for you to make them your own was always planned for and intentionally put there to inspire you to try improving them.

A lot of great development is iterative. We build a system, get it working, then go back and improve upon it. That's what this chapter is about, giving you ideas to go back through the projects and add new functions and features to rediscover them with the knowledge you gained since doing them the first time. This can be one of the challenges in coding. As we learn more or even just change tastes and preferences, we can look at a project and want to use totally different methods to achieve it. We've specifically tried to show you different methods and styles of coding with the projects in this series.

Let's run through each project and give you some examples of ideas and features that you could explore adding to the projects, but only in loose terms; it'll be up to you to figure out how to pull them off and make it your own. Even if you don't know the exact techniques for some of them, you should

DOI: 10.4324/9781003399018-11

have the familiarity and expertise to take on the challenge now. It's time to strike out on your own and conquer your own coding mountains.

Commenting

When we go back to our old projects, we can find them harder to understand than we imagined because we've changed our own thoughts and perceptions as we've grown. If we remember to think of coding as language, we can perhaps understand the change in more human terms, who as an adult still speaks like they did when they were a teenager? We learn new phrases, get new catchphrases, settle into new habits, accents, or mannerisms.

Because coding requires us to both translate and interpret as well as keep a mental model of the project or processes, it can be very difficult to read other peoples' code, and our own code if it's been too long since we last laid eyes on it. One habit we should get into, along with our students, is commenting.

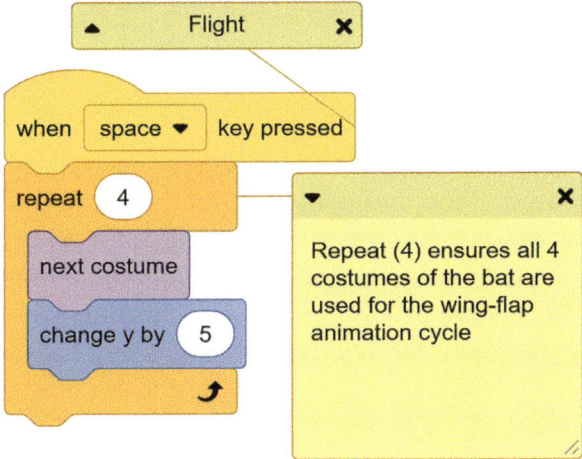

Figure 11.1 Comments can be collapsed (flight) or expanded. When expanded, they can be resized to any shape by dragging the bottom right corner. Switch their state by clicking on the arrow in the top left corner, or delete them by clicking the top right "x".

Figure 11.2 The four beginner projects covered in this book.

Commenting is the practice of providing text hints and descriptions added to a project to explain things using natural human language. Comments in Scratch take the form of sticky notes, either at large or attached to specific code blocks. Simply right-click to add a comment on the background for an at-large comment, or on a code block for an attached comment. You can resize comment notes as needed by clicking and dragging their bottom right corner. You can see I used comments to label all the stacks in my projects, but this is just one minor use. Good practice would be to be more descriptive and especially mention any connections with other code or objects.

Try to go back through the projects and add comments to describe what stacks do, what variables are for, what objects will react to what broadcasts, etc. Add notes to your projects that let you read and easily understand its structure, flow, and processes at a glance. This is a key skill for professional programmers and helps immensely when two or more people are handling the same project. Getting used to doing it will greatly improve students' skills while making evaluating their work easier as well.

Dinosaur Dance Party

Our first project is pretty basic when looking back at it now. You could try having the active dinosaur take centre stage while performing solos and then return when they aren't. You could add some more band members: for example, having seven dinosaurs instead of playing premade music, you could turn them into a basic music scale, each playing a note when a key is pressed. You could also add in more of a light show, with button presses to get specific light show effects. You could make the light show either a constant functioning animation cycle while selected or just a single, onetime event.

Fireworks Display

The Fireworks Display project was specifically created as an open-ended template, allowing users to take and adapt it for their own designs. The project works by pressing a button to launch a firework, but what about designing a sequence of multiple fireworks to shoot off when a button is pressed? Can you make a "fire all" button that works in addition to the "single shot" buttons? How about fireworks that shoot in different directions than just straight up? What about making the fireworks automatically reset after firing so they can be reused without resetting the whole project?

Batty Flaps

The original Flappy Bird game that this is based on was pretty minimal, but we can still come up with some fun ways to add to the game and switch things up. You could add collectables to score additional points by grabbing in addition to passing gates. You could incorporate sound effects. As well, you could give players awards based on their performance. Using what you learned in Butterfly Catcher, you could make sure the player can't flap when the game is over.

Butterfly Catcher

We saw in our Batty Flaps project how adding a menu can be both a nice visual effect as well as a convenient gameplay control. We could add a menu to our project with some other options. To change up gameplay, you could add in multiple butterflies, but have them vary in size and point value. You could also try re-theming the game with a different scene and use something other than insects for collectables/dangers.

Intermediate and Advanced Skills

We'll revisit these projects in the follow-up chapters in both *Book 2: Intermediate* and *Book 3: Advanced* as we improve our skills. We can use the skills and techniques learned in the later books to once again revisit these early projects and twist and expand them with new features and ideas. Be sure to check out the two other books in this series to learn more and take these and other projects to new levels!

12

Troubleshooting Scratch

Perhaps nothing strikes fear into teachers told to integrate coding into their classroom more than the thought of dealing with bugs, errors, and computer trouble. Admittedly, coding can throw a lot of surprises at you. Earlier, we even said bugs are a part of the process. This might not be very reassuring talk, but just like coding itself, we can prepare ourselves for these eventualities. Here's some advice for the most common problems faced in the classroom when teaching with Scratch, in an attempt to arm you with the knowledge and practice to overcome most of the potential issues you'll face.

Demonstrating and working through bugs and errors with students is a fundamental part of working with technology. We want to show them resilience, determination, logic, and best practices for dealing with setbacks, and that's true with any subject, be it digital or analogue. By taking the time to make bugs and errors part of our teaching, we can use them to touch on a lot of key skills and make opportunity out of crisis. Working through bugs can build confidence and resilience, teach solution-finding practices, help practice analytical thinking, and provide for social learning opportunities with students helping each other with brainstorming, analysis, planning, and helpful solutions.

Site Issues

If you're using the website version of Scratch (as opposed to the offline downloaded and installed program version of Scratch), there can be some

DOI: 10.4324/9781003399018-12

connectivity, browser, or website issues. That's without considering all the computer and connection issues – computers not booting, students unable to log in to their profiles, forgotten passwords, computers deciding now's the time to install updates, Wi-Fi not connecting, etc., which are all local issues with machines and networks. We are not dealing with those issues because they'll be specific to your school and school board, so hopefully you've been through all those issues with your local tech support. Instead, let's focus on issues particular to Scratch, assuming you've successfully reached the website.

Problem: The project is freezing/not responsive.
Possible Solutions:

1. Refresh the website (click the Refresh button in the browser).
2. Does the project use clones? If so, check if too many clones are being generated (over 300 can overload the website). Clones, repeats, or My Blocks generating clones is a common cause of this issue.
3. Close and restart the browser.
4. Reset the Internet connection.
5. Give it time (website may be temporarily down or overloaded).

Problem: Can't remix projects.
Solution: Log in to a Scratch account.

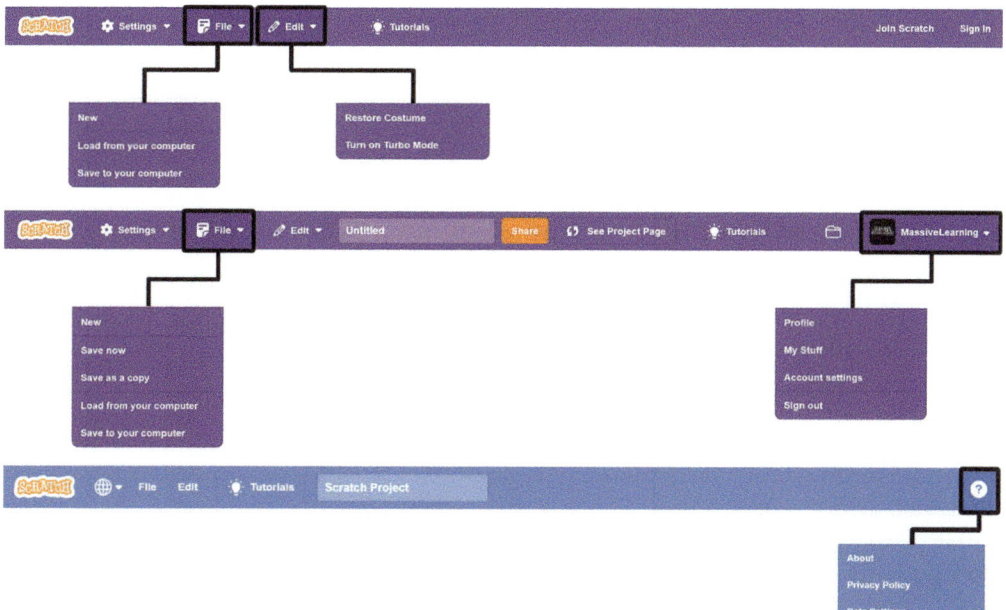

Figure 12.1 The Status Bar's File options when not logged in (a guest) or when logged in. Cloud saves and sharing are not available unless a user is logged in!

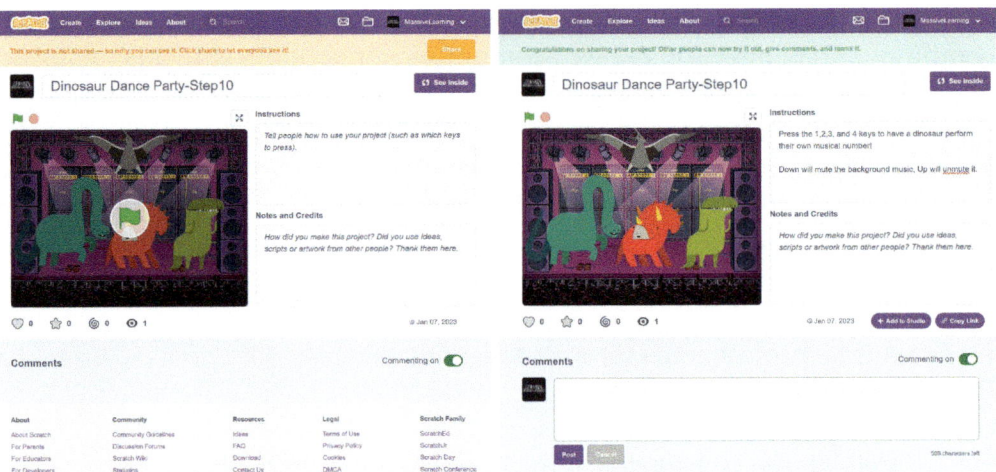

Figure 12.2 You don't have to share what you create, but it does help build the community. For education, it can allow teachers to share starter projects, or for students to do peer review or submit projects. Here's a project before and after sharing.

Problem: Can't autosave or save.
Solution: Log in to a Scratch account.

Problem: Can't name projects.
Solution: Log in to a Scratch account.

Problem: Can't access/save to/see the Backpack.
Solution: Log in to a Scratch account.

Problem: Can't access previous projects.
Solution: Log in to a Scratch account.

Problem: Can't find/see the project that was sent.
Solution: Project owner must ensure the project is shared.

Problem: Link doesn't work.
Solution: Project owner must ensure the project is shared.

Coding Issues

Most errors and issues you face will be coding problems. Computers are very literal in their understanding, and this is something most humans are not very good at. Our brains are built to predict, imagine, and leap to conclusions to act quickly and save time and energy. We often make assumptions or leaps that computers can't and, making our code, end up doing something we don't expect as a consequence of it. There are many forms of error and many reasons, but the fundamental process of thinking being different between humans and computers will likely always cause friction and bugs. The most irritating truth of bugs is, no matter how unwanted the behaviour is, the computer is only doing what we tell it to do.

While the site issues we listed earlier are relatively mechanical – and you probably already deal with those kinds of issues regularly – coding issues are a whole new can of worms. Most importantly, we want to prepare ourselves for how we think and deal with bugs and errors before we even start fixing them. A bug or error is, like in any other subject, an opportunity for reflection. When a student says something is going wrong, we don't want to rush to fix it for them. Can they identify the problem? Can they enunciate it? Can they solve it themselves? Can they eliminate some of the possible causes? Our first goal should be to challenge the student to overcome the obstacle, or at least give it their best shot. If they engage and try those things, we still don't have to leap in ourselves to fix it. We can then challenge the class to think about the problem, discuss it, and suggest solutions. This helps take the problem and turn it into opportunity for practice and class engagement. If that fails, then we can lead our own review, analysis, and solution to the problem instead of just solving it for that one student and ignoring the rest of the class. By having already engaged them in the problem, our time spent fixing the problem is helping inform all of them.

Now that we have some better idea of how to approach bugs, what are we dealing with? There are four basic areas where things go wrong.

The Wrong Object

These errors happen because the wrong object was selected. Either it's displaying the wrong information to the student or the student has assigned code or properties to the wrong object.

> **Problem:** "I can't find that code block" or "That code block isn't there" – the list of code blocks is different than expected.
> **Solution:** The student has selected the stage instead of a sprite. The stage doesn't have a number of the code blocks available to it since it can't move or do a number of other things that sprites can. Make sure the student selects the correct sprite first.

> **Problem:** The wrong object acts or reacts.
> **Solution:** Code was placed in the wrong object. Check the object that did act/react and the one that didn't, and you'll likely see some code assigned to the "wrong" object.

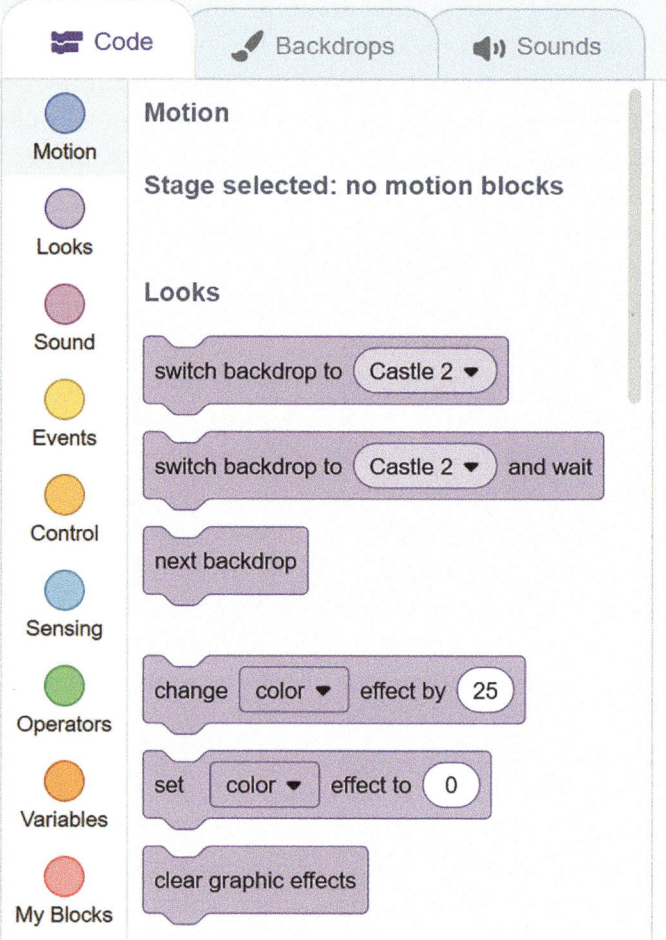

Figure 12.3 When the stage is selected, code blocks that don't apply (such as all motion code blocks) don't appear.

Alternate Solution: The wrong broadcast event might have been called, making an unexpected reaction happen. If the behaviour you saw is supposed to happen at a different time, that would indicate a wrong event instead of a wrong object issue.

Alternate Solution: When using a ●**((Property) of (Object))** driven behaviour like ●**[Move to X: (●(X Position) of (ObjectA)) Y: (0)]**, if you selected the wrong object, you could be referencing the wrong object's properties and making it look like one object is calling the behaviour rather than the other. Double-check any ●**((Property) of (Object))** code blocks and see if they are correct.

General Tips

Transferring code blocks. Remember all the ways to copy and transfer code blocks (drag, copy + paste, the Backpack). You don't have to rebuild things from zero. Save time by copying code that's incorrectly in one object over to the one that actually needs it to save time (but remember to delete the incorrect code after copying!).

Specify Objects. While instructing, be sure to specify what object you're coding for and when you switch objects. Repeating yourself to the class can be very useful for avoiding these mistakes.

The Wrong Block

There are a number of blocks whose similar nature can easily trip up new coders. These errors can also be hard for people to spot because of our habit to read what we expect rather than what's there. Knowing the likely culprits, you can look for them easier and quicker, as well as predicting the issue and making sure you clarify the difference when they come up in instructed learning.

Confused Pair: Left vs. Right

The turn code blocks for left and right turns can easily be confused, especially for younger students still uncertain with their directions. It's unlikely to cause serious issues, but if something is veering off in an unexpected direction and you used a right- or left-turn code block, check if there was a switch.

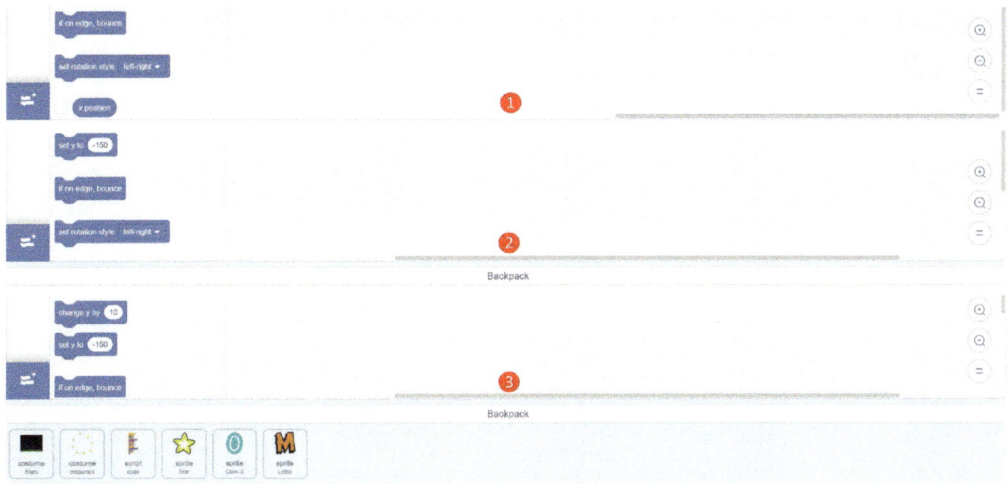

Figure 12.4 You must be logged in to access the Backpack. ① shows none. ② shows the minimized Backpack. ③ shows the Backpack maximized with some costumes, scripts, and sprites already added.

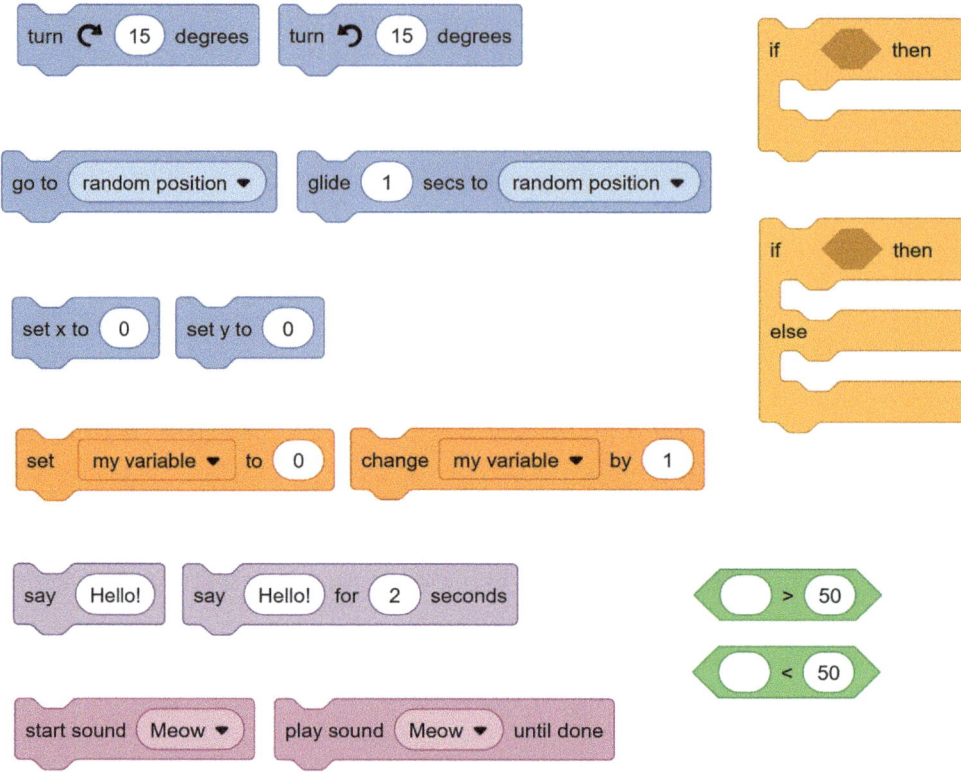

Figure 12.5 The most likely code blocks students will mix up. If your project uses one of these, expect somebody to grab its twin!

Sometimes, seeing the difference can actually be fun and interesting, like in our Pen Tool Fun project.

Confused Pair: Go To vs. Glide To

Both of these code blocks can send a sprite to a specific X/Y or sprite location, but they work differently. ●[Go To] automatically teleports a sprite instantly, changing its X and Y properties. ●[Glide] works by changing the X and Y properties over a set amount of time. Kids love glide because it animates, and they can see the sprite moving (gliding) to the destination, but it can cause some issues if used in place of ●[Go To]. Glide includes a time value, and this acts as a hold or delay on the code. Until the glide is complete, the computer won't move past that code block, which can cause some issues. If you want a sprite to move with glide and to detect collisions, you can't use both in the same stack, because while the glide is happening, it won't run the collision detection code because it hangs on that single timed code block. ●[Go To] can cause its own problems, as it changes instantly, so sprites look like they're teleporting and might be too sudden a change; you could even teleport past an obstacle without ever colliding with it like you expected.

Confused Pair: X vs. Y

It can take a while for X- and Y-coordinates to sink in. Even adults who have learned them can forget when they haven't been using them. When things move left/right or up/down unexpectedly, you can always double-check any X- or Y-affecting or powered code blocks. In the Properties panel, the X and Y properties have appropriate arrows beside them as a reminder of which is which, but there's also plenty of mnemonics you can introduce your class to (X is a cross/X is across, or Y points down, etc).

Confused Pair: Set vs. Change

There's a lot of features with set and change versions of code blocks: X/Y positions, size, graphic effects, volume, sound effects, •variables, and pen properties. Until students get used to working with variables, it's very easy getting confused between the two. Set makes a property/variable an exact number, regardless of what it was (absolute). Change adjusts a property/variable from its current position (relative). This difference between absolute and relative changes can be hard to describe, but projects can be a great way to visualize the difference, and it can be well worth taking the time to illustrate the difference with a visual example like position or size.

Confused Pair: Say/Think vs. Say/Think For

All four of these code blocks allow displaying some text on the screen. The •[Say] code blocks use a speech bubble, while the •[Think] code blocks use a thought bubble. The difference is "For", where a value is introduced, allowing a time scale that makes the displayed text disappear when it is up. Without "For", the text is permanent and only disappears with a •[Hide] or new •[Say]/•[Think] code block. The permanent versions are good for non-action, non-animating scenes, information displays rather than reactions or conversations. One needs to be more involved in ensuring they disappear when you need them to. "For" versions are great for fast-paced programs, where you want to maintain some action and activity, and you can set and forget them since they'll disappear on their own. Be careful the information displayed doesn't disappear too quickly for slower-reading users.

Confused Pair: Play Sound vs. Start Sound

Similar to •[Go To] vs •[Glide], the difference between these two blocks is time. A •[Play (*sound*) Until Done] code block plays a sound effect, but it holds the program on that code block until the sound effect has finished playing. A •[**Start Sound (***sound***)**] code block begins playing a sound effect but doesn't hold the program (other things can happen while the sound is playing), and code execution continues on without any delay. So •[**Play (***sound***) Until Done**] is useful for things like alerts or dialogue, where you want to pause things until it completes, whereas •[**Start Sound (***sound***)**] is great for sound effects and other incidental sounds that may even stack up and simultaneously play while other things happen.

Confused Pair: If vs. If/Else

This pair confusion won't tend to last long with the block structures limiting coding options, but I thought it is worth mentioning. It can be hard for younger kids to see an •[**If <True> Then**] code block and realize it isn't the one they want, so you need to be very clear about selecting the right one with younger audiences. The difference is, of course, the "Else" clause. An •[**If**] only tests a condition and allows code if the condition is found true. In an • [**If/Else**], the condition test determines which of the two clauses of contained code is used, the first if the condition is true, the second if not.

Confused Pair: > vs. <

This will be no surprise to teachers, coding experience or not. The •Operators blocks for numeric comparison can be confusing to younger students. Again, the key is clear communication beforehand to ensure the correct selection in the first place. We can use all the math class mnemonics to try to get the difference clear with our students. The nice thing about code is that with the computer responding immediately to the code, students can test and see the results on their own.

The Wrong Order

Even with the right sprite and the right code blocks, things can go wrong. Our third category of errors are "wrong order" issues, where code blocks aren't in the right sequence to execute as desired. This is where we start getting into more complicated issues. Thanks to Scratch's code block system with distinct shapes and colours for code blocks, a number of issues are much more visible and easily avoided. This visual feedback doesn't just help students follow and build things but can also help avoid or correct bugs.

Simple Sequences Code Flow

The most basic errors in this category are simply putting a code block above or below where it belongs. In many cases, the exact order doesn't matter. For example, in giving multiple variable values to initiate a program or setting multiple properties to represent a new state. In these cases, the exact order rarely matters; the properties are all set in a single frame of animation and don't interfere or influence each other. But in other cases, the order is extremely important. Take for example a sprite given a •**[Move (100) Steps]** then •**[Turn Right (45) Degrees]** versus a sprite given a •**[Turn Right (45) Degrees]** then •**[Move (100) Steps]** instructions. They'll end up facing the same direction, but almost as far from each other as they are from their starting positions.

Control Structures and Code Flow

As we introduce control structures like •**If** statements or •**Broadcast** events, we change how the code flows. Thinking of code through flow charts or process systems, errors can occur because we took a wrong turn in our chart. This could happen because we called the wrong •**Broadcast** event or were listening to the wrong one, triggering the wrong code to execute out of sequence. It could occur because we changed a control like a game state that triggered an •**If** statement at the wrong time and changed the expected sequence of events or a property at the wrong time. •**Control** structures are very powerful and handy, but a "typo" of grabbing the wrong block or value can have us swerve off course in the flow of processes.

Logic Clauses and Chains

Learning to use logic can be a challenging task; there are lots of errors to be made when including or excluding conditions in code. It can be hard to know

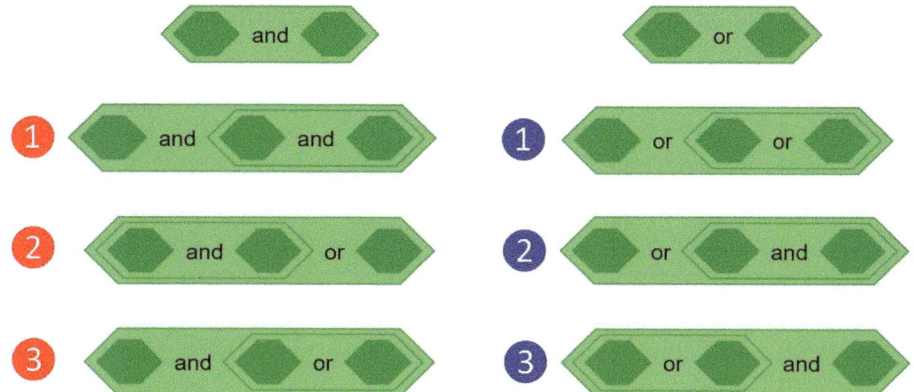

Figure 12.6 Combining logic blocks can easily become confusing. It's important to remember that how they are connected may affect the results. ❶ Ands can be combined and simply add to the necessary conditions to meet. ❶ Ors can be added to add additional options to meet. When mixed, the outer code block is dominant. So in ❷, you must have both the first conditions or have the third. In ❸, you must have the first and one of either of the second and third conditions. In ❷, you can have the first condition or both of the second and third. In ❸, you must have one of the first or second conditions plus the third.

whether to write some conditions inclusive or exclusive, whether to use an "If", or an "If/Else", or which two clauses to put your code in. If something strange is happening, always check the logic conditions; it's surprisingly easy to end up with the reverse of what you intended because of our love of leaping to conclusions, leaving the cold mechanical reasoning of computers in the dust. One very easy issue to fall into is chaining logic conditions – when you put multiple logic operators together for a single test. **<<<A>or <B>> and <C>>** will give different results than **<<A> or <<B> and <C>>>**. It can be very hard to spot this kind of error, so know to look for it!

Concurrency and Race Conditions

We can run into other problems in the flow of our code processes separate from those previously mentioned. Sometimes issues happen when we need to handle multiple processes that happen concurrently, or with (near) simultaneous execution of code. This can cause specific problems with order of execution. If any properties or variables are affected in one process that affects the other, the order of operations could drastically change the results. It can be challenging to make simultaneous systems work without improperly interfering with each other. When we don't know which process will complete first and they can influence each other, we call that a "race condition". The faster process "wins" and makes changes that could affect the other. Sometimes, the order can be predicted, sometimes not, so these issues can be very tricky. We can try to put code into single stacks or split it into multiple stacks to simplify processes to a linear run or to ensure concurrent execution as needed. Single stacks make the flow much easier to analyze and predict, but sometimes concurrency is exactly what you need. As we mentioned earlier, take for example the ●[Glide] code blocks: if an object is gliding and you need it to be clicked or collided with, you'll need to put the collision or clicking code in a concurrent process because other code in that stack won't execute until the glide is completed. Delay blocks may or may not help simplify these issues. Concurrency can be a blessing and a curse and usually very hard for the human brain to predict accurately. If something isn't happening, or it's happening at the wrong time, check for any block with a time value or time-sensitive word like "until", "for", or "done" and think about how they might be holding things up.

Other Errors

These three categories merely represent the most common issues a coding educator has to deal with. In this category, we'll deal with some of the other less-expected issues that can happen in Scratch that don't necessarily fit into a convenient overarching category.

Layers

Some of the simplest issues come from layering. Every graphic on the screen is printed in layers, each object laid on top of the objects below it, all on top of the backdrop. At times, this layering may cause issues. Large objects may completely obscure smaller objects underneath them. You may need to control the depth of objects at times, but the •[Go To [Front]], •[Go To [Back]], •[Go [Back] (1) Layer], and •[Go [Forward] (1) Layer] code blocks can be used to solve these issues.

Visibility

There are a number of issues you may run into dealing with object visibility. The first and most obvious is having objects disappearing. If you ever •[Hide] an object, you will need to have a •[Show] code block somewhere to make it visible, or you'll never see it again after it first disappears. While invisible, or hidden, a sprite has limited functionality. A hidden object cannot •[Say] or •[Think]. A hidden object also doesn't collide with other sprites.

Colour Selection

Scratch uses a graphical technique called anti-aliasing. This is a process that blends (and changes) colours to smooth out lines to appear more visually appealing and avoid the jaggedness that pixel designs and displays can cause. This looks great, but it can lead to a few tricky issues. If you are using any colour collisions, watch out for anti-aliasing blending the colours you think are involved. If you want to use a colour and use the colour picker, you may see that there are many more colours than you expected in the zoomed view of the screen. Any colour near a line or change of colours may end up being blended by anti-aliasing. This means that colour collisions looking for that colour won't find it because it changed to a different colour. This is especially likely to happen with sprites that are being displayed at less than 100% size or at anything but 0% ghost effect, which blends them with whatever is behind them to create the transparency effect. Colour collisions are great, but if you're having unexpected results, it may be because of these issues.

Clones vs. Originals

As we mentioned in our projects, clones can be tricky. Whenever you're using clones, remember that there's a difference between the original sprite and any clones you make. The original sprite will never receive the •[When I Start As A Clone] event, and the clones will never receive the •[When ▷ Clicked] event. Don't expect the original to act like a clone, and don't expect clones to act like the original. It can be hard to get used to it, but once you do, it is a very powerful system to work with. Also, remember that clones always start with the same properties as the original. If you don't see clones, is it because you had the original ●[Hide]? Unless you ●[Show] the clones, they'll remain hidden.

Wrong Concepts

The hardest of bugs is when we simply had the wrong conception of a system or method. Our brains ran off with an idea that wasn't true, doesn't work that way, or otherwise doesn't line up with reality. There's no predicting this and no preventing it. Sometimes we have ideas that don't work. We want to try to make them work and account for any bugs in the system, but sometimes the way we conceived something is simply not how it will actually work. Maybe we didn't account for some factor that makes it unable to work, or it is simply inefficient or impractical. It happens. Even to the pros. The way to address this issue is always approaching coding with a flexible mindset. There are many ways to address an issue, and sometimes we need to explore and try some options in order to really understand the problem or our tools in order to come up with better ideas. Learning can sometimes mean failing. Learning that something doesn't work, or work in that way, or work in this condition, is still learning. Learning those limitations and finding ways to work around them is perhaps the greatest part of learning coding.

Backup Plans

Having been a coding educator for years, I've seen just about everything go wrong. The worst issues are large-scale technological issues out of your control: the Internet is not working, power outages, websites go down, etc. Most of the time, educators can simply switch tracks and work on something not so technologically dependent, but for those of us that are coding specialists, how can we deal with the wrenches that sometimes fall into the gears? Here are some ideas about how to work around dealing with major technological hurdles and still deliver some educational opportunities.

Offline Scratch

One of the most impressive things about Scratch is that they have both the online instant access website and a downloadable offline version. If you have issues with spotty or slow connections, you can download it and install as an app that requires no Internet connection to work. This can be a wonderful tool for rural and remote communities, to ensure lack of Internet doesn't lead to lack of learning access.

Some educators may actually prefer the offline version of Scratch. By having the app, students won't be distracted by the shared content on the Scratch website. While I think the inspirational factor of the sharing platform side of Scratch is a great benefit to students, some may benefit from less distractions.

Another factor going for the offline version of Scratch is that it means students will save and sort their projects in their student profiles. This does mean the classroom organizing and viewing options aren't available but for prolific creators, but the ability to sort projects into folders in a way that personal accounts don't allow for can be a big benefit as dozens or even hundreds of projects pile up.

Most school board computers will have program installation locked down (and for good reason), so if you want to work with offline Scratch, you'll need to have the tech department install it on the student computers. The nice thing is, this can mean having someone else do the work and possibly automate the process, so it shouldn't be any additional hassle for teachers to have this capability in their classroom.

Pseudo-Coding

Let's be clear: "pen-and-paper" coding is absolutely no replacement for working with computers, but learning to do pseudo-coding can be a powerful and useful skill to add to the mix of tech skills we develop in our students. Pseudo-coding is when a coder writes out their design, not using the exact code word for word, but in a shorthand to give the broad strokes of their plan on how to achieve or organize things. It's more about planning than syntax and details.

Pseudo-coding can look different for different people, as each person makes plans or notes in their own way. In general, pseudo-code will help define the components of a project – such as the objects, backdrops, costumes, sounds, writing needed, as well as the processes, the events, and the stacks that will make things happen. Pseudo-coding can be as verbose or as brief as needed; plans can be quick sketches or fleshed out fully to the point of actual code, but often it consists of shorthand sentences, like "If near [objective] then change to warning state". Maybe they don't know what they want the warning state to be, or maybe they do, or maybe they have some idea about a feature like "Repeat 10 {size +10%, wait, size -10%}" as a note to make something pulse and a method to achieve it. You'll note the exact name of code blocks might not be listed, but the intent is clear.

In addition to more written pseudo-code, there can be very different methods for keeping these notes, as the creative process works differently for different people. Some may use more text document format, others may want to use spreadsheets to list ideas, some may prefer physical whiteboards, others might use presentation software to make things more visual, and flow charts could be used, mind maps, or even wikis for very large projects. No perfect system exists for pseudo-coding, creative designing, or visioning, since everyone and every project is different. Explore and experiment with different methods and tools, see what you like, and help your students explore all the ways their ideas can be brought to life. There are so many wonderful ways to be creative, and such wonderful things to create. You and your students have beautiful, thoughtful, meaningful, and fun treasures hidden inside them; by discovering and sharing the tools and opportunities to bring them to light, we all benefit. We hope this book has helped you learn, grow, and prepare to be a guide on that process and help usher in a beautiful era for human creativity and potential.

13

The Next Step in Your Coding Journey

Now that you've completed our beginner-level training, don't feel like you're on your own! We've got more training to support you. In Book 2, we cover off intermediate Scratch. This book will help take those who have developed a basic familiarity and comfort with Scratch through large and more complex projects, where we'll explore a lot more coding techniques and skills.

Book 2: Intermediate has four projects that are a step up from what we've done so far – they pack in a lot more punch! This does mean you should have a good familiarity with Scratch, since they don't hand-hold on basic tasks like finding different code blocks, but hopefully this book has gotten you there. With the basics already down, this allows us to explore more of what coding can do. When you're comfortable and confident with the projects in this book, have troubleshot any problems you've had, and have dabbled around with some other projects, why not keep moving forward and check out this next book?

In it we'll deal with larger projects and cover off dealing with multiple scenes and ways to transition between them. Our Pen Tool Fun project will introduce extensions and the ability to create art live in your project while its running. Our Interactive Story will show you how to create choose-your-own-adventure-style stories. The Snowball Fight game will show you some simple physics modelling and how to have player turns. The Big Map Racing will show you how to create tight, reactive controls, incorporate power-ups, and have gameplay over a map far larger than the screen. We'll have games with multiple levels and show techniques for changing levels and loading

DOI: 10.4324/9781003399018-13

Figure 13.1 The four intermediate projects we cover in Book 2.

Figure 13.2 The three books in *The Teacher's Guide to Scratch* series

different setups for them. We'll deal with complex systems like using game states or state machines, do some basic physics work, work with storytelling, state tracking, and even try a Scratch extension that adds new code blocks to work with in your project.

The intermediate Scratch projects we present will be ideal for students in grades 5 to 9. They should prepare you and your students to push the limits of Scratch and, by the end, should have them ready to advance onward to advanced Scratch in grades 8 to 10.

Beginner Scratch was just the beginning, but with the rest of the book series, we've still got your back. We're here to help you understand and incorporate coding, and we've got plenty more support to give you. When you're ready for it, join us for the next chapter in your coding journey!

14

Final Thoughts

So that's it for now. Hopefully, you've enjoyed the book and feel confident about using it with your classroom. There will always be more to learn and try, but if you've gotten through the four projects, you're well on your way and should know enough to handle introducing Scratch to your students. Scratch is an amazing tool, and with it you'll be able to unleash your students' potential as well as your own.

You've now done the hardest part of incorporating coding into your classroom – getting started. It might have been frustrating, obtuse, or confusing at times, but you made it. Use that memory to help guide your students through their own frustrations and confusion so they can get over the same hurdle of getting started. We don't know where they will take these lessons in their lives, but you've started them on that journey with even just the beginner lessons. You will have opened the door to a world of possibilities for them. Your goal isn't to help them finish learning to code – no one ever gets there – your goal was to give them the chance to see the path. We don't aim for 100% of students becoming programmers, but we need 100% of them to consider the possibilities of technology and become aware and thoughtful citizens. Even with these simple projects, you can do that.

Importantly, you're now a creator. Being able to make your own digital interactive media is an amazing skill to have. It's now your turn to challenge students to be creators in their own right. Help them express themselves and explore their interests and understand fundamental concepts, like logic, that govern all things. Interactive media is an amazing medium, and giving your

DOI: 10.4324/9781003399018-14

students the skill to create their own projects and share their ideas with the world is an amazing gift. We may not know what the future holds for them, but arming them with creativity, logic, empathy, and foresight is the best we can do to prepare them for it.

Through coding we can teach our students ways to think. Use it to reinforce their analytical abilities, planning, and ability to predict. By providing them with creative digital tools like Scratch, we teach them to play and experiment while pursuing a healthy, creative, and driven relationship with lifelong learning. Coding can be used for wonderful critical thinking practice, but also for empathizing through design. We can reinforce social learning and communication skills with group problem-solving and play-testing. Coding provides us a tool and a practice to refine and grow their thinking skills in a myriad of ways.

Coding may be a daunting challenge and, for many, something forced upon you, but it is also a wonderful opportunity. When we get over our fear of the new, we can dare to dream. We can explore a whole new world of creativity and thought and new ways to see and understand our world. We can provide our students one of the greatest sets of skills they can have for the 21st century and beyond. Through coding, we can teach our students to be creators and, with that, prepare them to create their futures and the world of tomorrow.

Glossary

⚐ – This icon is used to represent the green flag that is used as a start button for Scratch projects. It can refer to the start button in the Scratch editor above the Stage Window, or it can by in a ●[When ⚐ Clicked], which is the event that runs when a user clicks the green flag button.

✎ – This is the icon we use to represent the Pen tool extension in Scratch. The Pen extension, when added to a project, adds an additional category to the Code Block Library; instead of a coloured dot like other categories, it has a pen icon. When you see this icon, it means the following code block is in the Pen category.

.sb3 – This is the file extension for Scratch 3 saved files. Earlier versions of Scratch used. sb1 and. sb2. You can *"save to your computer"* and download an. sb3 file of a project and *"upload from your computer"* to load a file into Scratch.

Abstraction – This is one of many terms used to define computational thinking. It is the process of breaking a problem or goal into simplified and standardized components so simple and reusable processes can be developed to address those components. See also *computational thinking*.

Algorithm – An *algorithm* is simply a defined process to find a solution or achieve a goal. Math formulae are examples of simple algorithms. Recipes are another form of algorithm – a clear set of delineated instructions leading to a goal: food! Computer programs may be an algorithm or contain one or more algorithms.

Algorithm Building – This is one of many terms used to define *computational thinking*. It involves creating defined processes that, if followed, solve, or help solve, a portion of a defined problem or work toward a particular goal.

Argument – In computing, an *argument* is a piece of data passed to a function. Arguments are used by the function to evaluate a procedure or subroutine. They may also be referred to as *values* or *parameters*, though each has its own similar but distinct specific definition.

Art Workspace – In Scratch, it's the central portion of the editor when in the Costumes tab or Backdrops tab. It is comprised of the art tools and the canvas.

Asset – In general computing, an *asset* is something, not code, that a program uses to complete its function. This commonly includes graphics, sound files, 3D models, or data files.

Avatar – A person's representation on a digital platform. An icon, sprite, or even 3D model may serve as an avatar, depending on the context.

Backdrop – A graphic displayed by the stage in Scratch. The stage can have multiple backdrops, but only one can be displayed at any given time.

Backdrop List – When viewing the Backdrop tab in Scratch, the Backdrop List is displayed on the left-hand side of the screen. It shows all the current backdrops assigned to the stage that could be displayed and highlights the one currently selected.

Backdrop Tab – In Scratch, when the stage is selected, the second tab in the upper left-hand corner of the editor becomes the Backdrop tab. If clicked on, it provides access to an art workspace to create or edit backdrops and shows the backdrop list on the left hand side.

Backpack – In Scratch, if a user logs into their Scratch account, they can access the Backpack at the bottom of the editor. A small button labelled "Backpack" will show when collapsed, and clicking it expands to show a single row of thumbnails of what is currently in the Backpack. Users can drag scripts, sprites, backdrops, or costumes to the Backpack, where they will be saved and be able to be copied out from into any project the user opens. Items can be added to the Backpack by clicking and dragging them onto it. They can be deleted by right-clicking on them and choosing Delete on the pop-up.

Boolean Block – In Scratch, these code blocks are distinguished by their sharp, poky sides. They are used to provide arguments for some ●Control blocks, such as ●[If <condition> Then]. They are used to perform an evaluation that will return true or false, also known as a Boolean result. Most Boolean blocks are in the ●Operators category, but there are also Sensing category Boolean blocks and others. The ●<(0)=(0)> code block is an example of a Boolean block.

Broadcast – In Scratch, *broadcasts* can refer to a few different ●Event code blocks or to the unique named signals called Messages that are broadcast or received through them. Broadcasts are triggers that can allow code to trigger other code, including across multiple sprites or the stage.

C Block – In Scratch, numerous code blocks are shaped like a C or E, with gaps inside them that can fit other code blocks. These are called C blocks. The ●[If <condition> Then], ●[If <Condition> Then {} Else {}] are examples of C blocks. Code that is placed in the gaps of a C block is considered "nested" inside it in computer science terms. C blocks will exert some control on how the code placed inside them runs. Most C blocks are in the ●Control category.

Canvas – In Scratch, the *canvas* is part of the art workspace where you can actually draw and create images. The canvas has a centre point indicator and a line marking the 480 × 360 pixel area that will cover the entire Stage Window, but it extends past this area. In general computing, the term *canvas* can apply to anywhere you can draw or show images, but its exact use may be more precise depending on the platform/context.

Cap Block – In Scratch, any code block that ends a script (it does not have a bottom connector, so no code can go below it) is a cap block. The most common cap block is •[**Stop [All]**] in the •Control category. The •[**Forever {}**] code block is both a C block and a cap block because it repeats its nested code blocks endlessly and nothing can ever come after it.

Character – In computers, the term *character* can refer to either a typographic symbol, like a letter, number, or punctuation mark, or to a fictional being in a game, interactive novel, or other digital media. A character in a game could be a player character (PC), one that is controlled by a human player, or a non-player character (NPC) that is controlled by the computer. See also *player*.

Clone – In Scratch, a *clone* is a copy of a sprite. They function just like the sprite they are copying, but they never run the •[**When ▷ Clicked**] event since they are created while the project is running and therefor do not exist when that event is triggered. They do, however, get access to the •[**When I Start As A Clone**] event, which the original sprite they are duplicating will never be affected by. Clones are a very powerful technique to use in Scratch for populating a busy world or creating special effects, but Scratch is likely to crash if you create more than 300 clones at a time. Clones are dealt with through a trio of code blocks in the •Control category.

Cloud – In general computing, *cloud* systems are massively scaled and redundancy-protected systems to handle user data or even run programming. The online version of Scratch is a cloud system, saving user accounts and data on massive web servers that users access through the Internet.

Code Block – A compartmentalized instruction for a computer system and fundamental unit of block coding. Each code block has a specific function that it makes the computer perform. Connected together in the right order, they can create algorithms and programs, giving full complex instructions for the computer to fulfil a given series of tasks. Each code block's shape and colour will provide users with details about how they connect and what its function might be, or how to find it in Scratch's editor. Keep in mind: code blocks are not unique to Scratch, and there are many coding websites and programs that have adopted using them as the means to program.

Code Block Library – In Scratch, when the Code tab is selected, the Code Block Library is displayed along the left-hand side of the editor. It lists all the code blocks available to the user to use in their project. They can scroll up or down through the list, or along the very left-hand side of the editor, they can click on any of the categories (•Motion, •Looks, •Sounds, •Events, •Control, •Sensing, •Operators, •Variables, or •My Blocks) of code blocks to jump to that category. The extensions add new categories and code blocks to the Code Block Library for that project.

Code Tab – In Scratch, this tab in the upper left-hand corner of the editor gives users access to the coding workspace in the middle of the editor as well as the Code Library along the left-hand side of the editor. In the Code tab, users can access the code blocks and create code for sprites and the stage. To make art, the Costumes tab or Backdrops tab are selected. To make sound, there's the Sounds tab.

Coding workspace – In Scratch, when the Code tab is selected, the centre of the editor provides a workspace to do coding. Here, the user can arrange code blocks they drag from the Code Block Library to the left or by copying them in. The coding workspace is infinite, and users can click on blank areas of the workspace to drag it and change their view, creating more room as their projects grow. The code shown in the coding workspace is specific to the current selected sprite or stage. In the upper right-hand corner of the coding workspace, there will be a ghost image or watermark of the sprite or stage being coded for.

Collision Mask – In computing, a collision mask is a graphic that determines the positions that an object occupies and can collide or overlap with other objects. It may or may not match the size and shape of the object's display graphic. This method allows objects to have different visual looks compared to their collision behaviour, so in games objects can have artistic perspective that doesn't cause collisions despite literal graphic overlap.

Colour – In general, *colour* for digital systems is defined by three numbers assigned to red, green, and blue, or RGB. These control how bright each component light, affecting how a pixel on a digital screen looks, is turned on. If all three numbers are 0, the pixel is black, and if all three are maximum (generally 255) the pixel is white. Different combinations of values on each primary colour create different colours. In Scratch, *colour* refers to an exact output colour, what a pixel is coloured, but also refers to the hue of the final output colour, which is also affected by its saturation (how vivid or pale the output colour should be) and brightness (how bright or dark the output colour should be). It can also refer to an argument for the •Looks •[**Change [Graphical Effect] By (#)]** or •[**Set [Graphical Effect] By (#)]** code blocks that are used to shift the colouring of a sprite's costume, as opposed to the other possible graphical effects (like ghost or pixelate).

Colour Panel – The colour panel is a pop-up in Scratch that allows users to select a colour or fill pattern. It is used in numerous places wherever colour selecting is needed. In some places, some features of the colour panel are not available, depending on the context. Tips on working with the colour panel are in Chapter 4, "Scratch Basics".

Colour Picker – The colour picker is a tool accessed through the colour panel in Scratch represented by an eyedropper icon. It allows users to select a colour already in use, either in the canvas or the Stage Window, depending on the context. It will provide a zoomed-in view of an area and allow a single pixel to be clicked to provide that colour to the colour panel.

Combo – An unofficial term in block coding for whenever a code block is combined with another, especially when a Boolean or reporter code block, or code blocks, are placed into another code block. We also use the term *combo* to refer to some commonly connected code blocks, such as •[**When ▷ Clicked**] followed by a •[**Forever {}**].

Comment – In general computing, a *comment* is some writing in a program that is not seen by the user or interpreted by the computer but rather simply acts to inform anyone reading the code of something. Commenting code is a good practice that can help make it more readable by providing titles, declarations, explanations, and footnotes to the programming. In Scratch, comments can be added to either the coding workspace (by right-clicking on it) or to a specific code block (by right-clicking on it). Then, a sticky note graphic will appear. It can be moved, scaled, and typed into to provide a comment.

Computational Thinking – In general education, the concept of computational thinking helps define a system for problem-solving. It is comprised of decomposition, abstraction, algorithm building, pattern finding, modelling and simulation, and evaluation. It provides practical methods to solve problems through both analysis, planning, and execution. See also *decomposition*, *abstraction*, *algorithm building*, *pattern finding*, *modelling and simulation*, and *evaluation*.

Conditional – A *conditional* is a system of evaluation in computing. One or more metrics with one or more evaluations made of it returns a Boolean signalling if the condition is either true or false. For example, to check the weather in code could include •[**If** •<•(**Raining**) = (**True**)> **Then**]. Conditionals can also be numerical, as in •<•(**Temperature**) > (**32**)> or any other evaluation or comparison you might need on your project. In Scratch, conditionals are used with •Control category code blocks to make code run only if certain conditions are met so projects can be reactive and dynamic.

Costume – In Scratch, this is the graphic displayed to represent a sprite in the Stage Window.

Costume Library – In Scratch, if the user is in the Costumes tab, they can click the Add a Costume button to add a costume from the Costume Library. This lists all the graphics available by the default user.

Costume List – In Scratch, if the user is in the Costumes tab, all the costumes currently assigned to the selected sprite are shown on the left-hand side of the editor. The user can scroll up or down through the list. The selected costume displays in the art workspace. The user can click on any thumbnail in the Costume List to select it, and if desired, they can drag it up or down to change the order of the costumes in the Costume List.

Costumes Tab – In Scratch, the Costumes tab is available in the upper left-hand corner of the editor if the user has selected a sprite; otherwise, the stage selected and the Backdrops Tab will show in its place. Clicking on the Costumes tab allows the user to access the Costumes List and the art workspace to edit or create new costumes for the sprite they have selected.

Creator – In general, a *creator* is anyone using a program, or skill, to create something. In Scratch, it can mean the person using Scratch, the user of the editor, also known as a "Scratcher", or the person who created a project or asset that is being remixed. Because projects can be remixes of remixes or include assets from other projects or from outside Scratch, what exact relation a creator has to a project can be complex.

Decomposition – In education, *decomposition* is one of the components of computational thinking. It is based on breaking up a problem into distinct components to make smaller and easier tasks to deal with. This is useful in all problem-solving but particularly handy in coding because we can divide complex needs into discrete goals that can be easier to analyze and build solutions for, leading toward the end goal.

Define – One of the steps in the design thinking product design process. It comes after the empathise step and before the ideate step. It is where one takes the information and feedback from the empathy phase and identify a key problem and define it as an opportunity to brainstorm solutions to in the ideate phase. See *design thinking*.

Design Thinking – A design process model popular in the tech field. It is comprised of a cycle of five steps: empathize, define, ideate, prototype, evaluate/test. With product refinement achieved by rerunning the cycle. It reinforces the need of the user as the primary drive for design and ensures a strong link with the user through the design process. See also *empathize, define, ideate, prototype,* and *evaluating/testing*.

Direction – Sprites in Scratch have a direction property. This tracks or determines what angle they will move on if they ●**[Move (#) Steps]** and, depending on the rotation style property for the sprite, may orient their graphic to that degree angle as well. In Scratch, direction 0 is up, +90 is right, -90 is left. Sprites start with a default value of direction +90, or going right.

Duplicate – In Scratch, you can copy and paste code blocks, costumes, sprites, sounds, and backdrops. If you right-click on an object, you will get a pop-up of additional commands, generally allowing duplicating or deleting, but possibly others too. When you duplicate a code block, all code attached inside or under it will duplicate, too, and any typed or selected values will be copied as well. Duplicating a sprite will copy all its costumes, sounds, and code. There are many ways to copy things in Scratch. We give further explanation about it in Chapter 5, "Project 1: Dinosaur Dance Party".

Edge – In Scratch, the limit of the Stage Window view is called the edge. Objects collide with the edge the second any pixel of their costume touches it. The ●[If On Edge, Bounce] code block will make an object that touches an edge bounce off it using the correct reflective angle. In many cases of basic movement, a sprite cannot move past the edge of the Stage Window. It will only move so far as some portion of its sprite is still visible in the Stage Window. This protection protocol can be worked around using Set X/Y movement, but ●[Move (#) Steps] will be limited by it. The ●[Go To [*random position*]] code block may move a sprite to colliding with the edge but will never place a sprite beyond the edge outside of the Stage Window. See also *Stage Window*.

Editor – The Scratch editor (or just editor for short) allows a user to create a new Scratch project or edit an existing one. It comprises of the Stage Window, stage, Properties panel, Sprite List, and then the left side of the screen depends on the tab selected, providing a workspace and tools for that tab. While the Scratch website also has the main page, search pages, studio pages, project pages, and others, we deal almost exclusively with the editor in this book because that's where all the code and digital art creation takes place. In general computing, an *editor* is any program that allows users to create or edit files, such as a word processing program, and is not limited to coding, but it can also apply to other software for coding.

Empathize – One of the steps in the design thinking product design process. *Empathize* is the phase where designers work with potential user groups to hear their experiences, problems, and needs. This gathers the information needed for the next step, *define*, where those problems are refined as opportunities for development. See also *design thinking*.

Evaluating/Testing – One of the steps in the design thinking product design process. *Evaluating/testing* is the final step, occurring after the prototype stage, where a product or service has been created and rolled out. Evaluating/testing determines how the product or service performs in the real world. Products can be further refined and developed by repeating the cycle, going to another empathizing phase to gather feedback from users after the evaluating/testing phase. See also *design thinking*.

Evaluation – In education, this is one of the six components of computational thinking. It is used to determine the effectiveness of a process or solution and to see if it can be generalized and applied to other problems. See also *computational thinking*.

Event – In Scratch, ●Events are a category of code block. Most events are hat blocks. Events are generally triggers that determine when code will be run. Without being attached to a trigger, code blocks will not be run. See *hat block* or *trigger*.

Execute – In computing, to *execute* means to run code. The computer is executing the commands you have given to it. This can also be referred to as run or running. In some programming languages, code must first be compiled (converted to direct instructions for the computer at the lowest level) before executing. Compiling is not necessary in Scratch.

Follow – Because online Scratch extends beyond just an editor, providing a platform for sharing content, user accounts can follow each other. When following another user, you get updates about their activity, such as when they share a new project. See also *Scratch account*.

Frame – In visual art (including video, animation, and video games), a *frame* is a single set complete drawing of the screen – all the data that comprises what is shown on the screen, or window, for a project. This is like a single picture in a film reel, or a single composited cell of animation. A frame represents the view of the project, animation, movie, or game world at that exact moment. A game, animation, or movie will have a frame rate that determines how rapidly a new frame is composed and drawn to the screen to create animation and update the user's view of the project.

Frame Rate – In visual arts, this is how quickly the graphical view of the project is updated, given in a number of frames per second (or FPS). Most computer imaging is done at 30 fps, or 30 frames drawn per second, though modern gaming has moved to 60 fps as a common standard. Cinema tends to run at 24 to 30 fps. The human eye and brain are believed to have a limit of comprehension and perception in the 30–60 fps range. This means we might not reliably notice things faster than that.

Function – In general computing, a *function* is a clearly defined set of instructions for the computer to achieve some particular purpose. It can refer to a built-in capability of a programming language, such as Scratch's ●**[Go To [*Random Position*]]**, which automatically determines a randomized

X-coordinate in a range of -240 to 240, a randomized Y-coordinate in a range of -180 to 180, and then changes the sprite's X and Y properties to those new numbers. Functions can also be user-defined or custom, where the user creating a program can define a new function and provide all the instructions for it. This function can then be called in their code whenever needed, just like a prebuilt function can. The •My Block category in Scratch is the equivalent of building custom functions through the •[Define [My Block]] hat block and calling them through the custom stack block created for any •My Block that is created. Functions can have arguments or parameters that ensure needed numbers for the function to work are provided for it when it is called.

Grid – All computer graphical systems use graphs or grids to track the position and scale of objects they need to draw (show on screen). In Scratch, a coordinate system is used for the same purpose. It has its origin at the direct centre of the Stage Window, known as X:0 Y:0, and will centre a sprite. The Stage Window is 480 × 360 pixels, which gives visible coordinates from X: -240 Y: -180 to X: 240 Y: 180. X increases to the right, while Y increases to the top. Other platforms and digital systems may use different grid systems. One of the backdrops called "XY-grid" in the Backdrops Library can be used to see a diagram of the grid system in Scratch.

Hat Block – In Scratch, *hat blocks* are a shape of code block. They have a bump on the top that indicates nothing can attach above them. Hat blocks are all triggers, determining when the code attached below them should run. Most hat blocks are in the •Events category. Every script must start with a hat block to tell the computer when that code should be run. Without a hat block, the computer will never run the code. See also *event* or *trigger*.

Ideate – This is one of the steps in the design thinking design process. It follows the define phase, where the challenges or problems have been identified and defined as opportunities. *Ideate* is where possible solutions are conceptualized to meet the opportunities. After ideate, the prototype phase builds systems imagined by the ideate phase. See also *design thinking*.

Input – In general computing, input is feedback from the user or environment. It can be direct commands, such as key presses or mouse clicks; indirect commands, such as mouse movement; or passive input, such as sensor readings or clock ticks. Input is used to interact with the computer system, generally triggering some code or providing values or data.

Lag – In general computing, this slang is used to refer to delays in processes. Most commonly, it refers to any pause, disruption to Internet streaming for games, videos, or video conferencing. It can also refer to delays in stimulus–response systems or delays between decisions and results. Scratch can lag if you have too many clones running overly complicated code.

Language – In general computing, this can refer to either a user's human language or a programming language, a system developed to allow humans to give instructions to the computer that is more readable and convenient than giving more direct and literal machine instructions (also known as assembly). In Scratch, *language* refers to user language. In the editor, you can change the language Scratch uses by clicking on the "Settings" button in the upper left-hand corner of the screen. Scratch is available in 74 languages at the time of writing this.

Message – In Scratch, a *message* is part of the broadcast system. Messages are uniquely named user-defined triggers that can be used through three of the ●Event category code blocks. Messages are broadcast – a sprite broadcasts the message using the ●**[Broadcast [Message]]** code block, and then any and all sprites or stage with ●**[When I Receive [Message]]** code blocks will trigger and run their attached code. This is a very useful system to get multiple objects to respond to a specific event.

Modelling and Simulation – In education, modelling and simulation is one of the six components of computational thinking. It's about using models and simulations as both a method of understanding situations and processes as well as creating or changing situations and processes. Coding is an excellent way to utilize this method, as we can not only create models and simulations but also can run easily and efficiently even at massive scales, from which we can derive results, but we can also easily and affordably run enough simulations to derive statistical data for further analysis. See also *computational thinking*.

Nested – In general computing, nesting code is a method involved in certain functions where some code is run conditionally or in some way controlled by another function. It is contained by this controlling function and, in typed code, is generally indented from its controlling agent, which nests it inside, making it visually distinct in addition to its flow control aspect. In mathematics, we use parenthesis to nest certain math functions inside others to control the order of operations, similar to how nesting functions in coding. In Scratch, the C blocks nest code inside them, in some way controlling how the code inside them is run, such as conditionally with an ●**[If <*True*> Then]** code block or loop it like with a ●**[Forever]** code block.

Object – In general computing, objects are used in some programming languages as a model for how code is run and data is organized. An object is a discrete agent in the model, with properties that can be set or checked and have code run by it or affecting it. Infinite numbers of objects can exist in a model, and they can have different properties and behaviours. Scratch uses this system, with both sprites and the stage being an object. They each have properties, can have their own code, and can be affected by or affect others.

Parameter – In general computing, a *parameter* is a data point that helps define how a function should be run or is another form of data point that defines how the program, model, or simulation works. In Scratch, the Turbo Mode setting would be a parameter, as would the View Modes. The term is sometimes applied to any argument, value, or input that is passed to a function.

Pattern Finding – In education, this is one of the six components of computational thinking. It is also sometimes labelled as pattern recognition. It is the practice of analyzing data or systems to find patterns, either in behaviour or outcomes, or in similar processes and outputs. Pattern finding is often used to help standardize or simplify processes or solutions. See also *computational thinking*.

Pen Tool – In Scratch, the Pen tool is one of the optional extensions that can be added to Scratch 3, creating a new Pen category of code blocks. It provides the ability to draw lines or copy costumes to the background, on top of the current backdrop. The pen works by being set **[Pen Down]** (to draw) or **[Pen Up]** (to not draw) and then moved. The code is run by a sprite so that a sprite's movement will determine what is drawn if they have the **[Pen Down]**. If a sprite uses the **[Stamp]** Pen category code block, it will copy their current costume to the background.

Pixel – In general computing, a pixel is the smallest unit of graphical display, equivalent to a single point of colour. Computer monitors, TVs, and other digital displays are comprised of millions of pixels that display images by lighting up different colours. Digital art can be made in two different methods: pixel (defined by a grid of pixel colours) or vector (defined by mathematical instructions to construct shapes with given properties). All art is always displayed on the pixels of a digital display, though. In Scratch, the ●**[Move (#) Steps]** block will move an object approximately (#) pixels on the screen, though this can be rounded off because of the angle of movement (see *direction*). The Stage Window in Scratch is 480 × 360 pixels, though in Presentation Mode, it is enlarged, so 1 pixel in-game will be multiple pixels on your display. See also *vector* and *grid*.

Player – In gaming, a *player* can mean either any top-level agent in the game capable of winning or scoring or, more specifically, a human playing the game. Players can be represented with avatars, or through characters, which are fictional agents within the game controlled by players. You can think of players as actors, and characters as the role the player takes in the game world. See also *character*.

Project – In Scratch, users create projects as the distinct applications or programs they can run and share. Each unique project is listed in the user's Scratch account. Each project gets its own project page, may or may not be shared, and can be edited in the editor and remixed or seen inside if shared. See also *project page* and *Scratch account*.

Project Page – Each project created in the online version of Scratch gets its own website page on the Scratch website. This page allows the creator to add instructions and notes to the project and to share it. If shared, the page can collect likes, favourites, and comments and allow others to play it, see inside the project, or remix it for their own use. A user's Scratch account links to all the project pages for all the projects they've created and saved.

Properties Panel – In the Scratch editor, the Properties panel is directly below the Stage Window and above the Sprite Listing. This panel allows you to see and edit the most commonly needed properties of a sprite. It has dialogs for showing/editing the name, X position, Y position, visibility, size, and direction of sprites, though sprites do have more properties than those listed.

Property – In computer science, a *property* is a piece of data assigned or associated with an object. It defines some aspect of its state in the simulation. Sprites in Scratch have many properties: name, X position, Y position, visibility, size, direction, costume, and more. Coding allows a creator to assess and alter properties of objects in their simulation, making them move, change colour, appear, or disappear. See also *object*.

Prototype – This is one of the five steps in the design thinking design process. It comes after the ideate phase, where the conceptual idea for a solution or product has been created. The prototype phase is the actual building of the product or service to bring the ideate phase's idea to life. It is followed by the evaluation/testing phase, where the prototype is tested in the real world to see how it performs. See also *design thinking*.

Remix – All projects in Scratch are under an open-source license as terms of use of the platform. This means that anything you create in Scratch can be shared with the world with no commercial use or restrictions on sharing. If you log in to a Scratch account, any project page you visit will have a Remix button on the top right corner. Remixing makes a copy of that project to your account that you can then edit and modify in any way you like, with the ability to share that project. This allows everyone on Scratch to learn from everyone else on Scratch since they can access the code they used to achieve their projects. We can all learn from everyone else through any shared project. See also *see inside*.

Reporter Block – In Scratch, reporter blocks are one of the shapes of code blocks. These are the round-edged pill-shaped blocks. They represent values, numbers, strings, or data. They allow a creator to refer to a data point somewhere in their project, such as ●(X Position) or ●([direction] of [sprite1]). They can also be typed-in data, such as ●((#) x (#)). Any of the white oval spaces in any other code block can be filled either through typing in a value or through using a reporter block.

Reticle – In the Art tab, the canvas has its centre point marked by a reticle, or target symbol. It can be important to align your costumes in reference to the reticle, as this is the point which a sprite is based. The X and Y properties will align directly with the reticle position of the costume. As well when rotating, the rotation of the sprite will be centred on the reticle. When moving vector shapes (or groups) on the canvas, they will automatically snap to the reticle if they are brought close enough to assist with properly aligning things.

Scratch – Scratch is a coding platform built with an emphasis on primary education and interactive media. It was developed by the Media Lab at MIT and is currently maintained by the Scratch Foundation. Scratch uses block coding to allow easy, friendly access for even young students. It is not just a platform for creation but also for sharing through the use of user accounts, studios, project pages, and a searchable listing of all shared projects created in Scratch. It is currently in its third iteration, Scratch 3. You can try Scratch at http://scratch.mit.edu/.

Scratch Account – The Scratch website is not just a platform for creation but also a platform for sharing. It has an account system for users to organize all their projects, save them on the cloud, and share them with the world. Scratch accounts are not required to use Scratch but are a great benefit to users as they provide a cloud saving for projects, an autosave feature (when logged in), and the opportunity to share projects with others.

Scratcher – A creator that uses Scratch can be referred to as a *Scratcher*.

Screen Refresh – In general computing terms, *screen refresh* refers to clearing the draw buffer in an operating system and rebuilding the graphical data to display to the user. Typically, this is done at every 30th of a second or more on a computer. In Scratch, "Run Without Screen Refresh" is an option available when creating a ●My Block. If checked, the code runs differently than usual and will not wait one frame between repeats and other similar delays in normal code. This can allow you to create custom functions with ●My Blocks that can rapidly process data, set up levels, or other things. See also *frame, frame rate*.

Script – In general computing, a *script* is a separate sequence of code that can be called as needed from any other code in a project and is a way to compartmentalize the reused functions required for a program. In Scratch, a *script* is any stack of connected code blocks. A script must start with a hat block so that the computer knows when to run that code due to a triggering event.

See Inside – In Scratch, every shared project can be viewed inside and out by other users. On any project page, there's a See Inside button in the upper right-hand corner. If clicked on, it will open the Scratch editor with a copy of the project loaded. Users can see all the code, sprites, costumes, backgrounds, and sounds that were used to make the project. This allows users to learn how to make any of the projects they explore, turning the site's ability to share into the ability to have millions of users teach one another.

Shared – Projects in Scratch can either be shared or not shared by their creator. A shared project can be seen by other users on Scratch, who can find it in search results, visit its project page, and even see inside the project or remix it. An unshared project cannot be seen by anyone except the user that created it.

Size – In Scratch, all sprites have a size property that determines how large they display in the Stage Window. Size is a percentile, with 100 indicating normal size relative to its current costume's size, 50 indicating half size, and 200 indicating double size. Costumes can be checked for their base scale in the Costume tab's art workspace.

Sound Library – In Scratch, when the user has selected the Sound tab in the editor, the Sound Library can be accessed by clicking on the "Choose a Sound" button in the bottom left-hand corner. The Sound Library is a collection of music, notes, and sound effects available to users to use built-in to Scratch. In the Sound Library, sounds are listed as tiles with a name and a play button that if the user hovers their mouse over it will play the sound effect. In addition, at the top of the Sound Library, there is a search bar and categories that will only list associated sounds. Clicking on a sound will add it to the currently selected sprite (or stage) in the project.

Sound Listing – In Scratch, when the user has selected the Sound tab in the editor, the Sound Listing is shown on the very left-hand side of the editor. It lists all the sounds currently added to the selected sprite (or stage). The user can select any of the sounds by clicking on them and can then edit or listen to them in the workspace. The user can click, drag, and reorder them in the Sound Listing.

Sounds Tab – In the Scratch editor, this is the third of the three tabs. It allows users to add sound effects and music to the currently selected sprite or stage or to edit already-added sounds. The workspace in the Sounds tab allows to play the sound clips as well as apply a number of different transforms or effects to either the whole sound or to portions of it by clicking on the sound wave to indicate starts and stops. The Sounds tab is explained in detail in Chapter 4.

Sprite – In Scratch, sprites are the main workhorse of projects. They are the individual objects that appear in the project through their costumes, take action through their associated code, and make sound with their associated sounds. Each sprite is its own discrete agent operating code-independently but can interact and react to the other sprites. All the sprites in a project are listed in the Sprite Listing, while the currently selected sprite displays its properties in the Properties panel. Only one sprite, or the stage, can be selected at a time, and whatever is selected is the target for any work done in the workspace, whether code, art, or sound. In general computing, the term *sprite* refers to a piece of pixel art that will be displayed to represent an object, rather than being an object in its own right with swappable costumes.

Sprite Library – In Scratch, if the user decides to add a new sprite, they can click on the Choose a Sprite button in the bottom right-hand corner of the editor. This will show the Sprite Library, a collection of already-drawn objects available to all users of Scratch. In the Sprite Library, if a user hovers their mouse over one of the tiles, they may see it animate, showing the multiple costumes associated with that chosen sprite, but not all sprites have multiple costumes, so not all will animate.

Sprite Listing – In Scratch, all the sprites that have been added to a project appear in the Sprite Listing. This is found under the Properties panel, which is under the Stage Window in the bottom right-hand corner of the editor. Each sprite in the game is listed here in tiles, with a thumbnail displaying the current costume active for that sprite. The user can right-click on a tile to duplicate it or delete it. They can also click and drag sprites to reorder them in the Sprite Listing. The currently selected sprite will be outlined in blue, and a trash bin icon will appear in the upper right-hand corner of the tile that can be clicked to delete it. Whatever sprite is currently selected will display its properties in the Properties panel above, as well as show its associated code, costumes, or sounds in the workspace to the left.

Stack – In Scratch, a *stack* is a set of interconnected code blocks. The official name for a stack is "script", but *stack* is very commonly used. See also *script*.

Stack Block – In Scratch, the stack block is the most common shape of code block. These are the basic rectangular code blocks. They can fit above or below other code blocks and often have a value space that the user can type in a number or text into or can use a reporter block to fill to have it populate with a dynamic value from the project. Stack blocks connect into scripts, and each script must start with a hat block in order for the computer to know when to run the code.

Stage – In Scratch, the *stage* is a special object that handles the background in projects. It can have its own code, graphics (called backdrops), and sounds. The stage has limited code blocks that can be assigned to it because it cannot move, change layering or size, among other limitations. It is always directly centred in the Stage Window and is always the exact same size. The stage cannot be removed or deleted from a project.

Stage Window – In Scratch, a Project is displayed to users through the Stage Window. This appears in both the project page and in the editor. The Stage Window is a view into the simulation created in Scratch, whatever form it has taken – game, music video, interactive story, etc. The stage will always provide the background of the Stage Window since neither the view the Stage Window provides nor the stage is able to move. The Stage Window is 480 pixels wide and 360 pixels tall. Its border is called the edge. Whenever the project is played (through clicking the ▷), all the action and interaction will occur in the Stage Window. See also *edge*, *grid*, *project*.

STEAM – In education, STEAM is an acronym standing for science, technology, engineering, art, and math. It represents the earlier term STEM with the addition of *arts* to highlight the need for creativity. STEAM and STEM are terms associate with a push to highlight knowledge of the physical sciences and applied science in both formal education and after-school programs or hobbies. Coding is often used as a way to incorporate technology and/or engineering into STEAM curriculum.

STEM – In education, STEM is an acronym for science, technology, engineering, and mathematics. It is a common term used to highlight a focus on physical and applied science in education. Coding education has often been used as a major focus for STEM programs as a way to incorporate both technology and engineering. STEM is sometimes extended to STEAM with the addition of *art* as an additional focus.

Step – In Scratch, a *step* is a measure of distance, roughly equivalent to 1 pixel. In computing, it will often be used as a reference to the execution of the sequential order of code, with each line of code (or code block in Scratch) equating one step of execution.

String – In general computing, a *string* refers to a sequence of characters or typographic symbols. Words, sentences, and passwords are all strings. They are a form of variable or data point that is not numeric – therefore cannot have math operations performed on it. See also *value* or *variable*.

Touching – In Scratch, sprites can be tested if they are touching other sprites or colours, or if colours are touching other colours. *Touching* in this case is determined by the costumes of sprites involved, or the sprite involved and the colour of the background, or any other sprites as composited into a frame at that time. It is determined if any pixel, or pixels, of the assigned colours overlap in position. In general computing, *touching* is known as colliding. Importantly, touching only counts overlapping, not adjacency.

Trigger – In general terms, a *trigger* is anything that sets into motion an action or reaction. In coding, this is often an input (such as a key press) from the user or an input from a sensor system. The hat blocks in Scratch are examples of triggers; they are specific events that will cause Scratch to recognize their occurrence and can then be used to have the computer run associated code assigned under them. See also *hat block, event,* or *input*.

User – In general terms, a user is anyone using an application, program, or project. This can refer to either the person playing a project in Scratch (who can also be referred to as a player) or as the person using Scratch to create a project (who can also be referred to as a creator or Scratcher). See also *creator* or *player*.

Values – In general computing, a *value* is a data point. It can be assigned to a variable or required for a function or found in a data structure. The term is used widely and freely to refer to any kind of data point and may refer to an input or parameter. In Scratch, a value is a white oval in a code block, a data point required for the code block to do its job, helping define or quantify its actions. Values can be represented through reporter blocks, and reporter blocks can fit into value places in code blocks. See also *input, parameter, variable,* or *reporter block*.

Variable – In general computing, a *variable* is a memory assignment for the computer. It creates a reference name to a piece of data that the computer will hold in memory, which can be changed or referred to at will. By making a general reference, this data can be referred to at any point or can be modified as needed. In some computer languages, variables have set data types – integer or string, for example; in others, they are dynamic. In Scratch, ●Variables are used the same as in general programming as fully dynamic data points, but it is also a category of code blocks that are used to work with ●Variables. To work with a variable in Scratch, one must first go to the ●Variables category and click on the Make a Variable button to create it first. See also *string*.

Vector – In general computing, *vector* art is one of two forms of art, the other being pixel (also known as raster). In pixel art, the canvas is defined as a grid of cells, with each cell representing a single pixel with an assigned colour. In vector art, the art is defined by mathematical formulae and instructions on how to build dynamic relationally positioned and proportioned shapes, lines, and spaces. Vector art, therefore, scales perfectly without any loss of quality and can have true curves and smoothness, unlike pixel art. It is, however, more difficult to create complex designs with. Scratch can create and edit both forms of art. It is highly recommended you try learning both and encourage students to do the same. We talk more about the two forms of digital art in Chapter 4. See also *pixel*.

View Mode – Scratch has multiple ways of presenting projects to users. By default, either on the editor or the project page, you will see the Stage Window at normal scale. In the editor, you can also choose between three view modes – default, compact (the Stage Window is half size, so you have more room to code), or presentation, also known as full screen. In Full Screen mode, the Stage Window will expand to fill as much room on your screen as possible. You can access the Default or Full Screen mode in either the editor or the project page. Full Screen mode can help students remember that they can't count on users being able to click on code blocks to execute them and must make full controls that don't rely on the user being able to manipulate code or sprites as an editor. See also *Stage Window*.

Workspace – In the Scratch editor, the workspace is the largest central part of the editor. Depending on which tab is selected, it is where the user can create their code, create or edit art for costumes or backdrops, or listen or edit sound. It will change nature with appropriate tools, depending on the tab selected. The content of the workspace will depend on the sprite or stage selected. See also *art workspace* or *coding workspace*.

X – In Scratch, X is both a position property of sprites and a dimension of the grid used to determine position and scale within the Stage Window. X measures the left–right positioning. X: 0 is the centre of the Stage Window, with X increasing to the right. The Stage Window is 480 pixels wide, making grid positions range from X: −240 to X: +240. See also *Y*, *grid*, or *Stage Window*.

Y – In Scratch, Y is both a position property of sprites and a dimension of the grid used to determine position and scale within the Stage Window. Y measures the up–down positioning. Y: 0 is the centre of the Stage Window, with Y increasing to the top. The Stage Window is 360 pixels tall, making grid positions range from Y: -180 to Y: +180. See also *X*, *grid*, or *Stage Window*.

Index